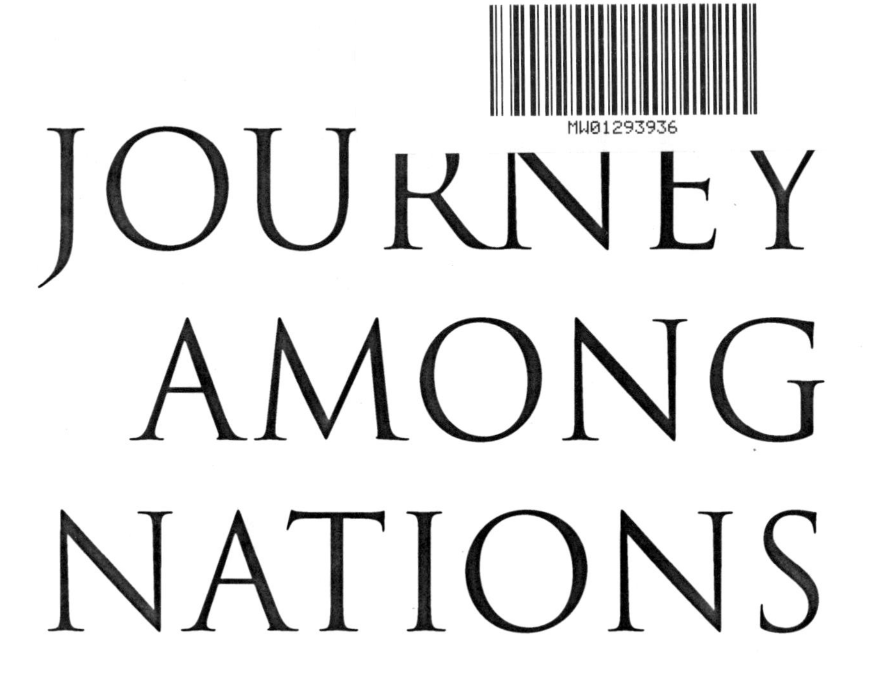

JOURNEY AMONG NATIONS

MICHAEL COHEN

Outskirts Press, Inc.
Denver, Colorado

v6.0 R1.0

Outskirts Press
http://www.outskirtspress.com

ISBN-10: 1-4327-0329-3
ISBN-13: 978-1-4327-0329-5

Library of Congress Control Number: 2007926984

Printed in the United States of America

A work of gratitude and thanks
to my supportive family
and loving wife

AFGHANISTAN
Keleft
Andkhvoy
Jeyretan
Mazari Sharif
Kholm
Rostaq
Fayzabad
Qala-I-Panjeh
Shebergha
Kunduz
Taluqan
Jorm
Eshkashem
BALKH
Khanabad
Dowlatabad
Baghlan
Farkhar
Sari Pul
Shulgarah
Samangan
Dowshi
Meymaneh
Tokzar
Qeysar
Hindu Kush
Khevak
Towraghondi
Mahmud Raqi
Nuristan
Charikar
Asadabad
Qala-I-Naw
Bamyan
Morghab
Mehtarlam
Karokh
Owbeh
Chaghcharan
Meydan Shahr
KABUL
Harirud
HIRAT
Dowlat Yar
Jalalabad
Puli Alam
Helmand
Gardez
Ghazni
Shindand
Khost
Qarah Bagh
Anar Darreh
Khas Uruzgan
Sharan
Harut
Farah
Now Zad
Tirin Kot
Arghandab
Kajaki
Farah
Delaram
Qalat
Tarnak
Khash
Lashkar Gah
Kandahar
Kadesh
Zaranj
Spin Buldak
Chehar Borjak
Helmand
Deh Shu
Gowd-e Zereh
National capital
Provincial capital
Town, village
Jewish Community
River/Stream
Main road
Secondary road

Afghanistan – *my land of origin – with its thirty million inhabitants is a marvelous country with a pleasant climate; deep snow in the winter and beautiful flowers in the spring. It is rich in stunning mountain and river views and is a world source of precious stones and natural gas.*

From the beginning of the 19th century, there was a kingdom that lasted until the 1970's. The lives of the people were quiet and happy. When a republic was established after a military coup, it quickly turned into an entity with extreme fundamental militias who led the country into anarchy and terrorism. The attempt of the previous U.S.S.R. to restore order totally failed, causing millions of victims, leaving casualties and refugees all over the world. Well-known terrorists found a home there...

On 9/11/2001, America knew a tragedy like never before. The subsequent deployment of troops into Afghanistan in search of the culprits ultimately led to the freedom the Afghans were so desperately longing for. With a new democracy in power today, we hope and pray for a peaceful and productive tomorrow.

ACKNOWLEDGMENTS

First, I would like to praise and thank G-d for all the good favors he bestowed on me until now, for all the good years during the Diaspora in my city of birth, Kabul in Afghanistan.

I am grateful to G-d who has helped me greatly for so long to live a quiet life in a foreign country which is a miracle by itself. As You helped me so far, I pray to You to help me further, to write this book to be a reminder to my children and descendants for generations to come. May the merit of my fathers and grandfathers stand in favor of me and my whole family, and may G-d fulfill all our good wishes.

I would like to thank my dear father, the jewel of my crown and my dear mother, a real righteous woman of valor, for all the advice and guidance which they gave me over and over so that I would choose the right way and understand matters properly. May G-d grant all their wishes, and bless them with long lives in happiness and good health. As well, to my whole family, friends, and colleagues who assisted me to write this book, which without their assistance this would have been impossible. Special thanks go to Haim and Avi Kohl who did their best to translate this novel in the spirit of its original version. Thanks to the DesignsDesigns.com team, Sharon Siegel for editing and formatting; and to Nachman Siegel for the graphics and cover design.

Please G-d, continue to bestow on me and all people who assist me, lives with grace and kindness and bring success in all our upcoming endeavors. And may we merit having a lot of satisfaction from all of our descendants.

CHAPTER 1

Afghanistan was a colony of the British Empire. In May 1919, the British pulled out their forces from the country after they lost a third consecutive battle against the Afghan people led by their king, Aman- Ollah Chan. The king retrieved all the powers of a ruler of a sovereign state and declared independence for the country. A huge celebration of this event was planned to take place in the capital Kabul. The palace sent out invitations to community leaders all over the country to come and attend this celebration. One of these invitations reached my Uncle Matityahu who was the newly-elected Kalantar of the Jews in the city of Hirat where my family originates.

I am very fortunate to know the story of the lives of my parents and grandparents, and am excited to have this opportunity. Tears fall down my cheeks while eagerly interviewing my mother and father. My parents have the *Mazal* to reach old age. My father is 97 years old and is very lucid. My mother is 88 with a phenomenal memory as in her youth.

My story begins as I was born to a modest and reputable Cohen family who date back many generations. Loving kindness has been a recognizable trait of our family. My father, *Mollah* Israel was born in 1911 in the city of Hirat to his parents, Yeshaiau and Esther, Z"L. When Israel was 2 years old, his mother passed away and he was raised by a nanny. My mother, Yocheved, the daughter of *Mollah* Yehuda Ha'Cohen, was born a decade after him.

I start the interview with this question, "Dad, what do you remember from your childhood and if possible, your earliest childhood memory?"

My Mother immediately interrupts the conversation, saying, "What could he remember from childhood if he doesn't remember what happened yesterday?"

Dad immediately responds with sarcasm, "When I was born, where were you? I still remember events from my early childhood!"

To prevent the interview from turning into a quarrel, I try to calm my father by telling my mother, "Wait a bit and your turn to talk will come too."

Dad happily continues: I remember very well when your mother was born. We lived and were raised in the same *havli* and we are cousins. I grew up without a mother, and my stepmother, Miriam, who was my late mother's sister, became my nanny. She raised me with a lot of warmth and attention. Miriam is also an aunt from your mother's side.

"What? Sounds a bit complicated?" Interrupts Mom and she explains, "This was customary at the time because of lack of choice. Marriages were done within the family."

I turned to my Father, "Please continue, where did your grandparents come from?"

"They came from Iran to the city of Hirat after being persecuted, and they liked the place because of its pleasant weather and the abundance of food they found there, compared to the famine that was present in other cities. They lived in a pervasive state of fear because of persecution by the Muslims and they needed to use nicknames. Why? For security purposes as in other Diaspora countries."

"What were their names?" I ask. "They were Cohen for many generations but they were called '*Ak Levi*'. I don't know their exact date of arrival, but I do know that they were among the first Jews who established the foundations of the city of Hirat."

"Mom, maybe you know a little bit more?"

"Yes, I can tell you the history from our parents' side. Great-grandfather Mordechai son of Elazar Ha'Cohen arrived at the age of four from the city of Yazed in Iran in the year 1875. From an early age he showed signs that he would be a rabbinical scholar, and later be a great rabbi and a preacher. People called him Rabbi Mordechai Babaee and he established many synagogues and schools. People enjoyed studying with him and listening to his Torah sermons. His wife, Great-grandmother Sarah was a righteous woman. Together they had sons and daughters: *Mollah* Yehuda, my father, his brother Isaac who died at a young age, and my uncle Matityahu who was elected to be the *Kalantar*, the head of the Jewish community in the city of Hirat. His daughters were: Esther, the mother of your father, Dinah, and Miriam who was the stepmother of your father.

"Mom, doesn't it sound complicated?"

"No, I don't think so. As I said in the beginning marriages were done within the family simply because of lack of choice."

"And what about your mother's side?"

"Also Cohen. Grandpa *Mollah* Yosef Ha'Cohen and his wife Rivka were very unique in kindness and fear of G-d."

"And how far back do you remember?" I asked Mom with curiosity.

"I remember when I was four years old, my brother Isachar was born and his circumcision ceremony took place in the synagogue. As the eldest daughter, Dad brought me from home to attend the ceremony. It was difficult for me to climb down the stairs and Dad helped me.

I do remember, as if in a dream that there was a revolution in the capital that was called *Bachahi Sakab*. Because of this revolution, there was a real danger that people would break into Jewish homes. Therefore, we took all our valuables, put them into a basket, and lowered it by rope into a *cha'a*. The Jews were instructed by the community leaders to lock their homes and to gather in the great synagogue. Dad took his hunting rifle with him out of fear and we hid in the synagogue for a few days until the danger passed. Thanks to G-d's kindness and due to the Jewish leaders led by Uncle *Mollah* Matityahu Cohen, friendly relations were established with the royal family and the heads of state."

As Dad heard the name of his uncle Matityahu he immediately interrupted, "I remember him as a tall, tough hero, and due to his courage and persistence, a friendly relationship was established between the Jews of Afghanistan and the king Aman Ollah Chan. In 1920, the king ordered all mayors and community leaders of his kingdom to come and appear before him in the capital Kabul on a specific date. At that time, to go to Kabul was dangerous, especially for a Jew. Just two years prior, a Jewish merchant, Akajan, son of Shlomo Ha'Cohen, disappeared while traveling from Hirat to Kabul carrying gold and precious stones. After a long, extensive search, his body was found somewhere in Kabul. This tragedy and the obvious risks did not deter Matityahu from his intentions to meet the king on time together with the other representatives. Prominent Jews and family members tried to convince him not to travel, but they failed. Left with no choice, they hired a bodyguard for him as well as a horse and supplies for the long journey and he joined the other mayors on their way to Kabul."

And Dad continues, "Before Matityahu left for his risky journey, all the Jews gathered to wish him farewell and they prayed for him to return safely. It took him 40 days to reach the capital and the king's palace. During the meeting ceremony, they called for each representative by name and when the king asked for the representative of the Jews, Matityahu jumped from his place, "I, your majesty. My name is Matityahu Cohen and I am the representative of all the Jews in the kingdom." "

Come here my son and stand near me," said the king. After a short introduction, the king placed a medal on his breast, seated Matityahu next to him and assured him of his protection. When the ceremony ended, the king told Matityahu that he may stay at the palace for as long as he wants and immediately assigned him a room and a personal guard. After a month in the palace, Matityahu decided to return to his family and again had to travel the risky journey back home. Only this time, he was very confident and felt like a hero who succeeded in his mission."

My Mom adds, "During his stay in Kabul, Matityahu succeeded

to establish close relations with the mayors and heads of state and on the day of his return to Hirat, the whole Jewish community welcomed him joyously and their enthusiasm increased when they saw him in his official uniform with the medal on his breast. Since then, he became the hero of the community. He also saved Jewish women who were charmed by Muslim men whom they ran away with and forced them to convert to Islam. Matityahu risked his life to smuggle away these women and send them to Israel via the city of Mash'had.

Suddenly, there was a moment of silence so I asked, "That's it? The end of the story?"

My father replied, "Really you think so, my *Pasar Jan*? Our story is only just beginning… Ask further."

"How many Jews lived in your neighborhood?"

"We numbered about four thousand Jews. All of us were concentrated in the same neighborhood. We had four synagogues: The first was the *Mollah* Yoav synagogue which was the most luxurious and was also called the Cohen-Ha (Cohenim). Second was *Mollah* Shmuel. The third was Golaikia. The fourth was called the Garji Synagogue. Besides this, we had two *Cham-Ula*, with about three hundred children. The rabbis and *Chalifa* were Asher Garji and his brother, *Mollah* Yecheskel Garji who was also a rabbi, a ritual slaughterer, and a circumciser. They taught Torah to the Jewish children. Later on the cantor and teacher, *Mollah* Yosef Bakshi, joined them. Even though he was blind, he personally knew each child. He knew the whole Torah and other holy books by heart and translated them into the local language. He also served as a cantor with a fantastic voice that the community loved to listen to, especially during the high holidays."

My Mom adds, "The community rabbis requested at every gathering and sermon that we all learn and keep to the Torah and most of all, to never forget our identity and the customs of our forefathers. And indeed, the whole community, with no exception, never changed their names, dress, and language. Everyone had names from the Bible and Jewish sages.

"Dad, which synagogue did you belong to?"

"We have been members of the great synagogue, Cohenim. As a practical man since youth, I was a caretaker, a messenger, and a public servant along with taking care of my dad and great-grandmother. Every day I walked long distances to accompany 10-15 children to the *Yeshiva* and back home. Since childhood, I loved to assist anyone that asked for help. At that time, a special event occurred in our community: the *Alliance* organization sent teacher and guides from abroad to teach our men, women, and children foreign languages, mathematics, and secular studies. The rabbis protested this idea which in their view was a danger to the religion and a catalyst to assimilation. They threatened that anyone who will register for secular studies would be expelled from the community.

"If so," I asked, "What was your daily curriculum as children? Just holy studies?"

And Dad answers, "Yes, we also played in the yards of the neighborhood, like one called *handak gunje*, a pit of treasures."

"Were there really treasures?" I ask.

Dad answers, "There was a Jew named Hanoch that always liked to rummage in this pit and we children used to throw things, mainly garbage into it, as a game. If Hanoch was there, and he noticed the children approach, he would shout at them to leave. After a while, there was a rumor that Hanoch found there gold treasures, took them with him, and traveled to Israel." Thus, several years elapsed.

Mom interjects, "Great-grandma Rivka reached an old age of 100 years old and she was still in excellent health. She always needed very little for herself. She was thin and small and loved to sit in her doorway and to prepare wool balls."

Dad adds, "Great-grandma was very modest and wore cloths that covered most of her body. The family respected her, listened to her advice and I used to help her out with all her needs.

Mom says, "I was 10 when great-grandma matched me to your Dad who was at that time 19."

I ask, "Has it been a successful match?"

Mom says, "I don't know… could we refuse?!"

"And what do you have to say about this, Dad?"

With a smile on his face he answers, "Certainly! Could she have gotten someone better than me?"

And Mom continues, "My Dad was then in Russia due to his businesses and did not know at all about this match. Great-grandma rose from her seat, noisily wished good luck, brought a package of *shiriny* and distributed it to the whole household. Everyone praised Mom that she was well-groomed and a good housewife, while Dad was praised as productive and diligent. Briefly, everyone accepted this match. My Father heard about the engagement after he returned from Russia and he angrily shouted, "How could you make such a decision without consulting with me? I am the father of this girl and I absolutely do not give my consent to this match!" Finally, great-grandma the matchmaker intervened and after using a variety of convincing arguments, Dad gave his consent. Dad then accepted him like his son and brought him into his business as a partner and was very satisfied with his work."

"If so, you and Mom must have gone out often, while she was still a child!"

"No! My father interrupts, "At that time it was customary that when a bachelor arrives at his fiancée's home, she would hide until he left. Everything was done with modesty and it was forbidden to leave the engaged man and his fiancée alone until their wedding.

I continue, "And what did you do with grandpa Yehuda, your future father-in-law?"

"We made business together with dried fruit, spices, furs and foreign currency exchange, which was common at the time. We managed business relations between Russia, Kabul, and other Afghan cities. Mom's father stayed more time at home while I traveled outside the city. Thus, my father-in-law had free time to study and teach the kids."

Mom says, "My dad stayed at home most of the time, especially on Saturdays. During the long winter nights, he taught us the stories of the Torah with maximum devotion."

I return to Dad with a question, "How was it to work together with your future father-in-law?"

"Very nice and satisfactory to both sides. We had a need to provide a living simultaneously for a few families. My own father was old and I was responsible for his livelihood. When I joined the businesses of my father-in-law, he was already disappointed with his current partners, named Haji Moshe Aaron and his sons, Sasson and Yosef. When the sons grew up and could manage businesses on their own it was decided to end the partnership. After checking the inventory and calculations, it was discovered that my father-in-law owed his partners a thousand rupees. It was a small amount indeed, but the bottom line was that he left the partnership empty-handed. He then decided to become a ritual slaughterer and chose *Mollah* Yecheskel Garji to teach him. After a few months he finished his studies and was sent to work in his new profession in a small Jewish community in the city of Kalla-Nov and its surroundings. For extra income, he made business in dried fruits but generally, his economical situation was not good. Therefore, he decided just to stay, save a little money, and return to his family."

Mom said "Since I was a first-born child, Dad wished to see me married. He wanted to return home at least with a little money for wedding expenses but finally, he returned empty-handed. Then he started a new partnership with Baruch Yekuthiel in the business of spices, dried fruit, and furs."

I turned to Dad, "And what did you do at that time?"

"I worked as a messenger and laborer for a rich Jew named Haji Moshe in Russia. After a year, I moved to Kabul to deal with dried fruit, furs, and foreign currency and I sent half of my earnings to my father, his household, and to old grandma. I was among the first Jews in the capital Kabul.

"How old were you then?"

"Twenty." He replied.

"And did you think about Mom?"

"Certainly! All the time I thought of the coming day that we could marry."

"And what about you, Mom?"

"The same. I was eleven years old but aware of being engaged. It was a period of poverty and we appreciated your father's assisting us from afar. He used to send us sacks of rice, dried fruit, and other food but especially money for daily living."

"Dad, one more question for you, How could you leave the whole family and your fiancée and move to a totally foreign city?"

"Why not? Everything went according to plan. First in Russia, by *Haji* Moshe Aharon and afterwards in Kabul which was a city of opportunities, as opposed to Hirat with its high unemployment. I met in the capital a smart merchant named *Mollah* Yosef Simantov who represented wealthy businessmen from Iran and Pakistan. He was a courageous man with good ties in the governmental offices and with the king Nadir Chan."

"How did he achieve this status?" I asked.

"Listen carefully, this is a long story indeed but I will try to make it short. Yosef Simantov knew the poor economic situation of the king and his administration and that they did not have money even for basic security purposes. They did not even have enough to pay the soldiers. Yosef donated money to the kingdom thus becoming a regular at the palace. Once, he discovered a plot to smuggle beautiful, green precious *zamarod* by rebels to Pakistan and he quickly reported it to the king. King Nadir Chan sent soldiers who captured the contraband. Right after, Yosef received a signed and sealed letter from the king which gave him privileges everywhere in the kingdom. For instance, he determined how much tax he wanted to pay for his imports. He always carried a bottle of anise liquor called *arak* in his pocket from which he would take sips whenever he wanted, even in the presence of the heads of government. Thus, he gave me a bottle to carry, so he could use it whenever he felt like. Besides privileges he received in his businesses, he also received free medical services of the highest level. Briefly, Yosef lived like a king. What a shame that the reign of the King Nadir Chan ended suddenly with his murder in 1933.

What a scary and depressing day it was for me, especially since Yosef was out of town and I was by myself in the store in the *sarai*!"

Continues Dad excitedly, "I did not know what to do at this time of chaos. In my possession was plenty of cash money divided into a few packages and there was danger from burglary and robbers. The supervisor of the sarai warned the business owners to lock their stores properly and to secure them during night hours. I hid the packages of money in various spots, and biggest one I placed behind a wall closet. We hired armed guards who were posted on the roofs during the night hours. In early daylight, we returned to work and to our good fortune, we saw that nothing was stolen. Meanwhile, Prince Zahir Chan became king.

"And what do you remember from those days, Mom?"

"In our city of Hirat, we heard about the chaos in the capital caused by the king's murder and great fear gripped the Jews. We hid in the Great Synagogue and thanked G-d we were saved." Mom continues, "I was already 16 and great-grandma asked over and over – Where is the groom, Israel? It's time for him to marry. How come no-one is calling him back? We then mailed out letters to Dad begging him to return."

Dad said. "I was already 26 and my boss, Yosef Simantov, urged me to return home." He said, "Why are you here while your fiancée is sitting in Hirat waiting for you? You saved 200,000 Afghan rupees which is a lot of money, and now you are capable of establishing a family. Listen to my advice and go back." And Dad continues, "I listened to their begging, tied the package of money behind my back, and left to my home in Hirat."

"Mom, did you know that Dad was returning with money and you were going to marry?"

"We didn't know that he was on his way back, but we did know the whole time that one day it would happen and we waited for him every day to come. When he returned, the whole Jewish community welcomed him joyfully and enthusiastically and right away scheduled the wedding – the 4th of November, 1937. Dad came back a wealthy man and decided to have a luxurious wedding. Besides family members, he invited all his friends and acquaintances and actually the whole Jewish community. Dad took care of all details starting with food purchases, dishware, and gifts for the bride and her family. On the wedding day, the women were very busy with cooking and organizing the event, while the community prepared themselves. The wedding ritual, managed by Rabbi Yecheskel Garji, took place within the Great Synagogue and right after, all the guests went outside for the wedding meal. An orchestra composed of a few musicians, headed by the singer/cantor, Yosef Bakshi, provided a good a time for the guests. There was a joyous atmosphere for everyone until late at night. Afterwards, seven days of celebration took place with seven blessings, for the bride and groom. Each day, and especially on Saturday, unique songs are sung, as customary in the Jewish religion.

Dad said, "In a feeling of euphoria along with a lot of attention and spoiling, I was committed to return after 40 days, to my place of work in the capital Kabul.

And then I asked Mom, "How would you describe the wedding?"

"Simply wonderful. When your father does something, he does it with all his heart and in the best way." She said.

"Dad had to return, why didn't you go with him? After all, a groom and a bride in their first year of marriage have to be together according to custom!"

"You are right. The family intervened and Dad decided to stay with me regardless of the outcome at his work. As usual, we relied on G-d who would always help us."

Left: King Aman-Ollah Chan (1919–1925) established peaceful relations with the Jewish Community.

Below: My uncle, Matatyahu Cohen was sent with a delegation in 1920 as representatives of the entire community. As a result, many consessions were offered to the Jewish community by the king and other heads of state.

Royal Palace built by King Aman-Ollah Chan in 1923 by French architects.

Top: My Maternal Grandparents, Mullah Yehuda and Sara Cohen.

Bottom: My Maternal Great-Grandparents, Mullah Yosef and Rivka Cohen.

My Parents, Mullah Israel and Yocheved Cohen.

King Muhammad Nadir-Chan (1925–1933) and his son King Muhammad Zahir-Shah. King Nadir-Chan was aided financially by several wealthy Jews during his short reign as king. He was assassinated in 1933 and succeeded by his son.

Mulla Haji moshe aharon

Mullah Yosef Simantov was a close friend to the king and was allowed into the palace at any time.

Mulla Mosa Gol and his business partner, Molla Haji Itzhak Basal, were also close ties with Royalty. They donated an airplane to the Royal family.

My Parents' — Mulla Israel and Yocheved Cohen — marriage contract. The wedding ritual, managed by Rabbi Yechezkel Garji and Rabbi Baruch Garji (Ed). In the presence of honorees Rabbi and Cantor.

November 4th 1937 city of Hirat Afghanistan.

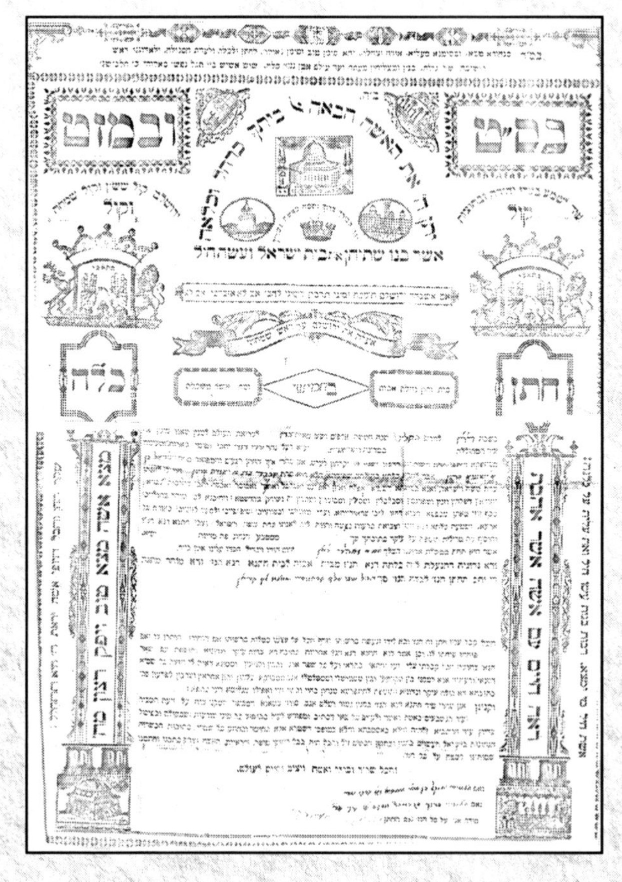

חתן

כלה

Rabbi Mulla Yechzkel Garji

Rabbi Baruch Garji (Ed-Kedoshim)

Rabbi Mulla Asher Garji

Mola Yosef Bachshi (Cantor)

Rabbi Shalom Bachshi

Rabbi Mulla Avraham Yakoteal

CHAPTER 2

Mom continues, "About a year after our marriage, Uncle Avraham, my grandfather's brother, Rabbi Mordechai Ha'Cohen, decided to travel to the land of Israel. Turkey, who ruled that territory for hundreds of years, retreated after World War I and England claimed the land. The Turks left behind chaos, poverty, and hunger. Suddenly, Uncle Avraham decided to immigrate to the Holy Land with his wife and three children. No-one dared to do this before or after him. There was no transportation like airplanes or cars that traveled such long distances and the path was full of danger. At that time we had not seen yet an airplane up close and when one passed over the city of Hirat, all the kids climbed onto the roofs to see the spectacular view of a 'tiny bird' making a great noise. Even for the adults, it was a curious and amusing spectacle. Cars were hardly seen then, and only a public vehicle would rarely pass the city. Trucks had been seen only in commercial centers and on highways. There is a story, that when the first time a small car entered a village called Hazara-jad, the villagers thought it was an animal walking on all fours and offered it straw to eat!

Uncle Avraham was getting ready, physically and mentally for his journey. He was full of fear indeed, but his heart drove him to the West, to the holy places of Jerusalem and its surroundings. The members of the Jewish community who saw how determined he was with his plan, encouraged him to implement it and he and his household started their journey. After a month of difficult traveling, they reached Tehran, the capital of Iran, via the city of Mash'had.

From there, they continued to Baghdad in Iraq. In Baghdad, they were delayed and tragedy befell them.

As told, one day, Uncle Avraham was sitting on the patio of the motel and recited psalms loudly and passionately. Fanatic Muslims noticed him, and thought he was a disturbed Jew who was performing magic. They invited him to their home with the excuse that he would pray for their success while their true intention was to kill him. Since then, he was never seen again despite his family's efforts to find him for many days. When their money ran out, they started begging in Baghdad in order to make enough money to continue their journey to the land of Israel. It was so terrible!"

"If so, then you must have had quite unpleasant adventures?" I asked my mother.

"Indeed! All that we passed was from G-d for bad or good. Listen and I will tell more."

"Mom, your past really interests me and I would listen to your stories with no end. I am even ready to write a whole book about the previous generations in our family, for the sake of future generations to know their history."

"It sounds clever," responded Mom and she returned to tell about her first year after marriage. "Dad stayed in Hirat and managed the business for his boss in Kabul. One day, we received good news that I was pregnant. The whole family was excited, especially grandpa who was going to have his first grandchild. Fortunately, I was surrounded by a close and supportive family, parents, brothers, and sisters who spoiled me. Dad supplied me with everything I needed even before I asked."

"And Dad, what do you remember from that period of pregnancy? You were going to be a father. Were you excited?"

"Certainly, it's natural. Even though I was very busy at work, I found time to spoil your mother to the maximum extent. For instance, I bought her a sewing machine and cloth in order to keep her busy with her beloved hobby, sewing. She enthusiastically sewed shirts, pants, and other types of clothing for the whole family, from great-grandpa to the children."

"Wow, how nice that must have been for you Mom! The most important thing was to make her happy while she was going to be a mother for the first time."

Silence followed, and then my Mother asks, "What else do you want to write about?"

"About everything, Mom."

"Yes, I just remembered something but it is about a different topic…"

"It doesn't matter; just tell me whatever you want."

"Okay, I don't remember the exact time of this event but it left me a deep impression. A member of our community named *Mollah* Akajan Basal hired two Muslims to clean the well that was the source of water for the whole neighborhood. Water from the deep well, which was in his backyard, was pulled up by bucket and rope. From time to time, it was necessary to clean the walls of the well from the dirt that had accumulated. The two laborers worked hard in cleaning when suddenly, one of them stumbled and fell into the well. *Mollah* Akajan immediately ran out calling for help and succeeded to recruit passersby for the rescue effort. It didn't help as the victim drowned and they barely succeeded to retrieve his corpse from the well. An outrage burst out among the Muslim neighbors and a complaint was filed to the police about a Jew employer intentionally killing his Muslim laborer. Policemen came to arrest the alleged murderer, *Mollah* Akajan, who foresaw such a scenario, and decided to take precautions. He offered the other laborer a huge amount of 1,000 rupees with the condition that he would tell the policemen the truth of what happened. The Muslim accepted his offer and went with the policemen directly to the courthouse to testify that his friend stumbled during work and the Jew was innocent. With G-d's help, the judges believed his testimony and *Mollah* Akajan was saved from a bitter fate."

Dad said "These cases were not uncommon and some of them ended in tragedy. *Mollah* Avraham Gad, a pleasant old Jew with a white beard, used to wander the *bazaar*. One time, on his way back home, he was ambushed by a Jew-hating man who hit him hard on the head. A pain-wracked *Mollah* Avraham responded instinctively. Why did you do that? What did I do to you? He responded: *Inshalla Ta Chana Narasi* (I wish you not reach home alive). After taking a few steps, the man collapsed dead in the middle of the street in front of passersby. Avraham, pain-filled and frightened, immediately rushed home and locked the door behind him. Rumors of the event spread quickly through the city and a huge group of Muslims came to demonstrate in front of his home. In a short while, they became excited, screamed, and proceeded to break down the door. They intended to drag him out and lynch him for causing a man's death. The Imam of the Muslims from the nearby mosque Masjid heard the commotion and rushed to the place asking, "What is going on here?" When the people reported to him what happened, he was shocked and ordered them to disperse immediately, since the accused Jew was a holy man, who if he stepped out from his home and cursed you, you would turn to corpses". Indeed, all of them ran away."

I responded, "This story is true, as I've heard it before a few years ago in New York from his grandson, *Mollah* David Gad." And then I turn to Dad with another question, "Did you really sense such a deep hatred from the Muslims?"

"No," answers Dad. "Most Muslims actually liked the Jews as you can see. We had the very best houses and businesses, land and luxurious *baq* and we mingled with the rest of the people freely and securely. The mayor and the managers of public offices related to us very nicely and the local police commander would regularly ask how we were doing and even asked us to submit him a weekly report of anyone or anything that bothered us. We had a steady supply of kosher food like fish and chicken that was brought by horses and donkeys; the private cars of those days in Afghanistan. Almost every household kept a horse and a donkey for travel and transport.

During the hot summer days, we went to the baq just for the weekends, while in the spring during the *rose sal,* we stayed from after *Pesach* until the eve of *Shavuot*. We used to load all the equipment and necessary food supplies on the donkey and traveled by foot to the baq which was just a forty minute walk. We took care to have enough people, ten men at least as customary, to pray together on Saturdays. Some rabbis used to go back home for the weekend in order to pray in their own synagogues."

Mom interrupts the conversation, "Yes, I'm recalling that once we prepared a lot of tasty food special for the holy Shabbat there. A few bowls for the evening meal and another two big bowls for breakfast which we put on a special fire-oven to keep warm during the Shabbat. We had a great and joyful evening, eating and singing Shabbat songs until late at night when we fell asleep. In the morning while approaching the fire-oven, we discovered that it was empty. The two big bowls with all our delicious Shabbat food, meat,eggs and fish, were stolen. What a fiasco! What should we do now? We looked all over and interrogated the laborers but nothing helped. We were forced to eat the remains of the evening meal. Such adventures were survived. Some of them were very weird, like the case of the wizard Mir Ali. I heard about him but I never saw him. Maybe your father did?"

"Yes," said my Father. "I remember him as a tall, dark-skinned man with a black beard who practiced magic and was surrounded by impure spirits. He used to come without invitation to Jewish homes disguised as a cat or dog and frighten the people there.

The families didn't like his hoaxes, and kicked him out. He madly took revenge. He would sneak quietly into their kitchens causing chaos, spilling food onto the floor, making the food non-kosher or switching big bowls of food between families. I would meet him wandering in the market and behaving perfectly normal. I even talked with him on several occasions and gave him money to calm him down. He needed money because he had big expenses for drugs like hashish, and he didn't bother anyone that helped him with money."

And Dad continues, "One day, he arrived at our home and old grandma greeted him with a plate of dried fruit. She made him swear not to come anymore because he scared the little children. He snatched the plate and we never saw him again. On the other hand, he did not stop visiting other homes in the neighborhood and especially the Bakshi family whom he abused as revenge, for them cursing him."

I asked Dad, "Did he steal anything?"

"No. He used to say that if he would have stolen, the spirit of impurity would leave him and the strength of his magic would vanish."

"Maybe we should change to a more pleasant topic such as the expected first child and grandchild. What were your feelings, Mom?"

"It was the most enjoyable period of my life. It was the year 1938 when I was seventeen and pampered by the whole family. "You are pregnant and you must strengthen yourself for the birth," the family told me over and over. I waited with little patience and I succeeded to pass the time by obsessively sewing on the machine that your father bought me. I sewed garments of all types for the whole family, especially for the expected newborn. Do not forget that back then there were no boutiques or clothing stores for babies. Finally, the ninth month of pregnancy came to pass and I was expected to give birth any day. Dad took care in advance for a midwife that would arrive the minute we called her."

"From where were you able to obtain a midwife?"

"Her son was a good friend of mine. She was an old Russian lady and served for many years as a chief nurse in a hospital." And Mom continues, "On Tuesday, the third of July 1939 at four p.m., I gave birth to a boy. The whole house was immediately filled with close relatives and women who came to help, only Dad was not at my side – he was on a shopping spree. One of the women removed her shoes and ran barefoot on the city streets to tell him the great news. Dad was euphoric! The whole community watched us since I was the firstborn of my family and this was the first grandchild for my parents. A boy, healthy and beautiful! What more could we ask for?"

Dad said, "I understood that the preparations for the *Brit* celebration would require much effort and I made myself ready emotionally and physically. The whole family stretched out a helping hand. The Brit was scheduled to be eight days after the birth on the coming Tuesday. Another Brit was scheduled for the same day for a newborn of a different family, the family of *Mollah* Reuven Amballu Cohen. Therefore, we decided to separate the two celebrations; ours right after the morning prayers and theirs in the afternoon. We bought a large sheep for meat as the main dish of the meals for the Brit celebration. For the dinner on the night before the Brit, we invited only close relatives and rabbis who also prayed and studied holy books in favor of our newborn. For the Brit ritual and the meal afterwards on the morning after, we invited the whole community to take part in our joyous celebration. We chose a young *mohel* named Chalifa Asher and the *Sandak* was *Mollah* Yeshaiau, Dad's father, as customary in our family for many generations. The boy was named Moshe Chai."

"Why?" I ask.

Mom answers, "This was the recommendation of my father, who said that as my name is Yocheved, like the name of the Biblical Moshe's mother, so the name Moshe would also fit our newborn child. I nursed him all through his infant days.

CHAPTER 3

"Three months after the *Brit*, Dad was called by his employer to return to Kabul. He stayed with us for about a year and a half, and his boss waited the whole time impatiently for his return. Eventually, his patience ended and Dad decided to listen to him."

I turn to Dad, "You had a big dilemma, of staying or leaving, but now besides separating from your wife, you also had a child to separate from."

"Yes, it was not an easy decision but I understood that especially now as a head of a family, I could not lose my place of work. In my mind, I thought that with time either they would join me in Kabul or I would return to them. I left my home and traveled to Kabul. Arriving there, I found plenty of work waiting for me and my boss, Yosef Simantov, was doing great in business and was reputable all over as a courageous man. I liked to work with him; he was very decisive and gave clear instructions. He liked me too and trusted my work as an honest man. He welcomed me as one of the family. I worked in the store in the *Sarai Shashzada* and did things for him no-one would have been capable of doing."

"Like what, I asked?"

"It's not so simple to describe it, but I will later."

I turn to Mom, "Were you missing the husband that left you with a baby to travel so far?

Immediately, Mom answers, "Always I missed him but plenty of work around the three-month old baby kept me busy and I was surrounded by all of my family, parents and siblings, who didn't leave me or the baby alone for one minute. You had to be there to believe what they did to this cute baby; they played with him like a toy non-stop. Everyone had a turn to hold him and be amused by him. I breast-fed and cared for him, and with time, he grew up to be a beautiful chubby boy with green-eyes. One day, my mother and I decided to take my son to a *hamam* to wash him. Many women from the neighborhood helped us to fill up the *taqare* with water. The baby really liked to play in the water and felt very comfortable to be washed. After we finished washing him, he didn't want to get out because he wanted to continue playing. He was only a year old but very smart for his age and understood what he wanted. Thus, the women in the wash-house said to me, "He is such a nice baby and wants to stay a little longer in the water and play. Leave him." My mother felt frightened and told me to take him out of the tub and go home. It was an irrational fear of the 'evil eye' of the women who stared at the baby too much, possibly, in her view, out of jealousy. Immediately, I took the baby out of the tub, wiped him and covered him with a towel. We then dressed him properly and left for home. The boy was tired and hungry so I nursed him and he fell asleep. Soon after, he became agitated and tearful and suddenly started to vomit. Then for a whole week, he refused to eat anything, even soft food or medicine drinks. It was also impossible to calm him down.

A physician was called and medicines were prescribed with no success. I recall that one of our close friends, *Mollah* Benjamin Chafi, knew an expert pediatrician and referred us to him. It was very difficult to get an urgent appointment but because of the emergency, the doctor decided to visit our home right away. By the time he arrived, the baby was pale and unresponsive. During the week, he lost three quarters of his weight and was very thin. The doctor determined that only a miracle could save him because his inner organs didn't function properly. Another week later, exactly on the same day and hour he was born, he passed away and his soul returned to heaven. We, the Jews used to say in such a case, "God gave and God took away. Blessed be He." We believed that his short life here was a reincarnation."

Pasar Jan, you had to see what went on then in my parents' home. Grandma was especially devastated. All of my brothers and sisters were not ready to accept such a terrible tragedy. When they took the child from my hands, I collapsed to the floor in sobs and I did not eat or sleep for a few weeks. How could it happen that such a child could be taken from me? On one hand I believed it was a decree from G-d but on the other I was not ready to accept it."

"Mom, how did you really overcome this tragedy?"

"My parents' belief in G-d was deeper than mine. They calmed me down by convincing me that it wasn't the end of the world and G-d will give me many more children. With this feeling I gradually overcame my grief."

"And where was Dad during all this drama? Did he know about this bitter tragedy?"

"No. He was in the capital city busy with his job. After we got back on track we sent a letter to the family of his boss, Yosef Simantov, to notify him of the disaster. The letter reached its destination and the son of the boss, *Mollah* Yehuda Simantov read it. He decided to delay notifying Dad because that week there was an important wedding in Kabul and he didn't want to ruin Dad's spirits."

"Dad, when did Yehuda tell you the bad news?"

"Soon after the wedding he called me aside and informed me about my personal tragedy. I was shocked and I withdrew into myself in sorrow. Friends were barely able to comfort me and after I recovered I decided to never return to Hirat. I wrote a letter to my wife asking her to join me in Kabul and stay with me there. Also, in this letter, I asked her father to accompany her on the trip."

"Yes my son, I received Dad's letter and your grandfather decided right away to accompany me and hired an aide for me; one of your cousins, a stepsister of your father, Amah Rebecca."

"What were you thinking at that moment? Was it a good idea to go to Kabul?"

"Yes. My father encouraged me. I was very scared and desperate and wanted a change after the tragedy. I was pale and thin, half of my regular weight, barely half of a human being. After all the preparations, my dad accompanied Rebecca and I on our long journey; partly on horse and wagon and partly by *lary*. On our way to the capital we reached the city of Kandahar where my dad knew a few merchants with whom he had business during good times. We stayed there for four days, sent your father a letter notifying him of our estimated arrival time in Kabul and continued to our destination."

"How did you feel, Mom, to leave your place of birth to a foreign city far from your close family?"

"Not concerned at all. After the tragedy with my son I thought I had nothing to lose. Besides, my father encouraged me all the way with his wisdom and guidance. I felt full with confidence and calmness especially because I knew that your father, whom I relied on very much, was expecting me in Kabul. I was aware of his strength to cope with difficulties and prevail."

"I turn to Dad: Were you happy that Mom was finally going to join you after almost a year?"

"Yes, I counted the days every day. It took them a long time to get there, over ten days. On the arrival day I asked one of my friends who had a jeep to come with me to the outskirts of Kabul, which was a one-and-a-half hour drive, and to wait for the newcomers. Before nightfall a lary arrived and my wife, her father and her aide – my stepsister Rebecca, stepped out and I welcomed them joyously. I helped them carry their luggage to the jeep and we moved immediately to my one-room apartment on the second floor in the havli where the Simantov family and other Jewish families resided."

Then Mom continued, "I thought that Dad had a house with a few rooms so we would have had more privacy. There I saw only one little room and I wondered how this would accommodate all of us. Your father encouraged me that this was only temporary and soon we would have a home with a few rooms. He told me that *Mollah* Yehuda Simantov and his family who lived on the ground floor of this apartment had intended to move out soon. He planned for us to replace them, but until then we had to get along with what we had. I remember well on the first night of our arrival, that Yehuda Simantov's wife turned to your father saying loudly, "What a beautiful wife you have. Wasn't it a waste that you lived a whole year by yourself? How could you do it?" She welcomed us with open arms and said, "You are very tired now. Go to sleep and we will talk tomorrow. Have a good night." We met her again on the next day and she offered to help us with anything we need. Then she invited us to a wedding that would take place in the neighborhood."

Dad corrects Mom, "No, the wedding had passed already. It was the last night of the "Seven Blessings" Celebrations which is no less than the wedding itself in the Jewish religion. We received an invitation to come as newcomers to the neighborhood."

"Then Mom said, "On the night of the celebration I got dressed with my best clothes and waited for Dad to come home and to take me to the event. At that moment, Avigail, *Mollah* Yosef Simantov's wife, entered our home with a big, beautiful shawl in her hand and said to me, "Yocheved, you are a beautiful woman. Put this shawl on your head and let's go to the celebration together."

Suddenly, Dad arrived in a good mood, dressed quickly and all of us went to the party. We were welcomed nicely by about sixty other guests. We already knew some of them from Hirat. They asked about our lives and where we came from and encouraged us on our new beginning. Then, someone turned to Dad and said, "You look so happy now."

"Certainly," Dad answered, "My wife has finally arrived and with her comes good luck. Now I pray and hope that with the help of G-d we will succeed. As we recite every Friday evening by song

> *Eshet Chayil Mi Imtza, Ve'rachok Mipninim Michra.*
> (A woman of accomplishment who can find, far beyond pearls is her value.)
> *Batach Ba Lev Ba'ala Ve'Shalal Lo Yechsar.*
> (Rely on her does the heart of her husband, and fortune he does not lack.)"

"Mom, how did you feel, at that moment, being in a new city among people you did not know?"

"It was indeed a change, but I already knew most of the people from before, except for some women who arrived from Buchara in south Russia. Anyhow, it did not take long to be friendly with them as if we knew them for a long time. I always made friendships easily, especially when we lived in the same havli. Within a short time we felt like one big family. Our best and closest friends were Apa Tzipor and her husband, *Mollah* Shimon Shamash. She was like a sister to me and loved us very much. Dad always helped her husband like he used to help other people. They were wealthy and respectful people. *Mollah* Shimon rented a house that belonged to the royal family. It was a huge house, surrounded by walls with huge gates. Within these walls were gardens and an artificial river like paradise."

"How could Shimon afford all this?"

Dad continues, "I was very close to him so I knew him a bit more. First of all, he was a good-looking man, always dressed like a European model with a suit, tie and hat. He spoke four languages and had a good friendship with the royal family. He was also in a business relationship with them involving jewelry and precious stones. Besides this, he was a foreign exchange currency dealer."

"Dad, why didn't you seek his help in business if he was so successful?"

"Shimon was a lucky man and we enjoyed being in his company, but while he and many of our friends were totally busy with making money in business, I was mainly busy and interested in being a public servant."

Then I ask him, "Didn't you need to make a living?"

"Yes, but in my view, when you help other people, a livelihood comes with it sooner or later."

"What type of kindness could bring a livelihood?"

"My son, listen to this story carefully and then you can judge for yourself. It happened during a period of *Farariha* by the Russian government, mainly Jews of Buchara and Samarqand. One day, on the eve of *Sukkos*, I was busy building a *Sukkah* for the whole neighborhood when suddenly someone came to me for help. He told me that three trucks loaded with immigrants from Russia have just arrived and need to be helped.

Without delay, I left everything and ran to welcome them. Then I saw plenty of men, women and children who had been persecuted by the Russians and ran away to Kabul. I thought to myself, where could I accommodate them? There were no hotels or hostels available. What should I do? I came up with a wonderful idea to direct them to the *sarai* and to turn all the warehouses into residences. I went to one *sarai* and the space was over-flooded with merchandise. Then I went to another and discovered the same thing. Suddenly, I recalled that a new sarai had recently finished construction. I traveled with the people there and met the owner who told me right away that my idea was possible. I found a few vacant lofts and immediately accommodated the people there.

Many of them, mainly the women and children, were on the verge of collapse. Then I ordered an immediate supply of food and wine for the upcoming *Sukkos* and quickly returned home to finish the building of the *Sukkah*. I didn't have much time left to begin the synagogue prayers for the holiday and therefore, I ran to the synagogue. Those Jews never forgot the good deed I did for them. Some of them succeeded tremendously later on in businesses in Kabul, like Avrashk Ahronov, Maidi, Avraham Polad, Tzion Moshpaqi and Sharbet."

CHAPTER 4

"The Jewish community grew and so did its necessities. We needed more rabbis to conduct weddings, ritual slaughterers for kosher meat, teachers for the yeshivot, and more. The need for rabbis was very urgent so we substituted them with circumcisers and ritual slaughterers, like *Mollah* Shalom Bakshi, *Mollah* Avraham Yekuthiel, *Chalifa Mollah* Yaakov, and more. The people of the Jewish community would come to me with the many problems that they had. One of these was match-making, which is a very interesting and complicated task."

"You were also a match-maker?" I asked Dad.

"Yes. The first match that I made was very difficult for me to accomplish. It was my first experience in this 'business' and G-d helped me out. The would-be couple, Shlomo Gad and the daughter of *Mollah* Moosa Gol, refused my matching twice and I needed a lot of effort to convince them that they fit each other. We the Jews used to say that "Match-making is as difficult as splitting the Red Sea." I realized how true this was. With time, I had become an expert in the field. When I saw an incompatible couple, I refused them right at the beginning. I had easy matches such as Reuven Balkhy with the daughter of *Mollah* Meir Ovadia who agreed to marry right at the first meeting."

"Dad, how did you succeed at such a difficult task?"

"With G-d's help, I succeeded to make about eighteen successful matches."

"Do you remember the names?"

"Almost all of them. After the two I already mentioned there were the matching of Avrashk Avraham Ahronov, Rachamim Maidi, Babajan Vardi, Matityahu Akshlomo, Meir Simantov, and Binyamin Basal with the daughters of *Mollah* Zion Raphael, Pinchas Amram, and more."

"Mom, what were you doing during the time while Dad was busy with helping the public?"

"I didn't see him much during the day because many people would approach him for assistance."

"Did you agree with and support him in this way of life?"

"Certainly, that's how we were raised, to be kind to people. It was a tradition in our families for many generations. People always came to our house since they were used to discussing personal problems with us and I was busy offering hospitality to these guests. There was a period when many immigrants cam to Kabul from Buchara, Balkh, and Hirat. In these cities, the unemployment rate was high and they came here to find work since business in the capital was blooming. Also, the medical facilities were on a high level and I recall many families that came from Balkh for medical treatment. They used to come for one to two months, recovered and left. Most of these families came to us first for help and I would pity them, cook for them, supply them with food and take them to the appropriate doctors. They looked miserable and many of them suffered from leprosy and large wounds covered their bodies, faces and hands; so how could I ignore them?"

"How did you do all this by yourself? You needed a lot of energy and finances."

"You are right, we were not rich people but I was always full of energy. *Pasar Jan*, our home was an open house. All you need is good-will, and then everything works itself out. Our close neighbors used to ask me, Mrs. Cohen, "How do you have so much patience and energy to cope with all these people's problems? Why don't you refer them to other places?" I always responded by saying that I have to do the job. Fortunately, Dad supported and assisted me in all these kind deeds we performed together. Besides, it was a good example for our descendants."

Dad suddenly recalled, “Once I was out of town in the city of Mazarsharif and when I came back home there were a few families from the city of Balkh and there was no room even for myself. My good friend, *Mollah* Shalom Shamash, used to say to me, “Israel, I see that many people come to you and always your home is over-flooded with guests. I have a few vacant rooms. Why don’t you refer some of them to me?” The next day, I turned to him for help in regards to these guests and he immediately offered them his rooms. *Mollah* Shalom was always ready to help others and especially in this period, the end of World War II, when many Jews from Russia and other cities in Afghanistan flowed into prosperous Kabul for work. I was mainly busy with two tasks: My work to give my household a living and second, giving public assistance using my good connections with the government.

Once upon a time, the leaders of the Jewish community of Hirat turned for help to the leaders of the Jewish community of Kabul, which I was a part of. Their problem was that the chief engineer of their city decided to demolish the Jewish cemetery of Hirat where renowned Rabbis and Hassidim were buried. This cemetery was well-kept for many generations when suddenly they received a sealed letter stating that on a specific date the demolishing process would take place because of a plan to build a highway in its place. We turned for help to government offices which repeatedly rejected us with the claim that the problem was in the jurisdiction of the city of Hirat and only they could take care of it. We then turned to high officials of the city of Hirat only to be informed that engineers from abroad came up with this plan and nothing could be done to cancel it.

In such a case, all that was left to do was to turn to G-d with prayers and beg to cancel this terrible decree. All the great rabbis together with their congregations gathered at the cemetery and pleaded to the sky with prayers and at the end asked forgiveness from the buried in case we didn’t do enough to save the cemetery. Afterwards, we left the place. A few days later, we received a sealed letter from the court stating that the whole plan had been canceled. This was a real miracle and this cemetery exists well-kept to this day.”

"Dad, it seems that you lived with the help of miracles, no? How could it happen that a Jewish minority in an Islamic country always made headlines in big businesses and were in connections with the royal family?"

"Listen my son: the Muslims in general and especially in Afghanistan believed due to experience, that Jews brought blessings to the country. The majority of the people was poor and had no education. We, the Jews, were educated and brought advanced methods in managing the country in all aspects of life. Actually, we created plenty of jobs in businesses and in construction for the Islamic population.

There was a time that a Jew couldn't establish a business on his name because the ministry of commerce and industry would not give him a *Jaz Chat*. The Jew would then look for a Muslim partner who usually was a good friend of his and known as an honest person to register the company on his name. The partnerships were not always balanced, but in accordance to the percentage of shares the Muslim acquired in the company.

This trick worked nicely and I do not remember any problems or frauds. The trust was mutual and they liked us because these businesses prospered. As we made money, we also spent it buying from the Muslims, huge quantities of food like fruits, vegetables, fish, and more; and mainly for the Shabbat and Jewish holidays. They were aware that we, the Jews, are good buyers and thus provided them a good living."

Mom interrupts on this topic, "We were used to having peddlers knock on our door in the middle of the day offering us different kinds of merchandise that they carried on donkeys such as *boteh* to fuel fireplaces and ovens. This boteh would accelerate the combustion of the wood. Other merchandise was *gandana, gil sarsui*, and more. We always needed these things. I remember a tall man who came to us every two weeks to sell sheep and goat milk for the cheese we made at home. These men always liked to enter our nice *havli* of prominent inhabitants, offering various types of merchandise, since they knew that the Jews are good buyers.

We lived in a spacious and beautiful house in this *havli* surrounded with good neighbors, like the whole Shamash family: *Mollah* Shimon Shamash, his wife, Tzipor, *Mollah* Chaim Shamash, his wife, Toti and their son Meir, who was the son-in-law of *Mollah* Shalom Shamash. Later on, *Mollah* Chasid Shamash and his wife, Dora joined us in the *havli* and we became very close friends with them.

I recall when a friend gave birth to her fifth daughter. Her husband was so frustrated after having five daughters. He didn't want to go to the hospital to visit his wife; even though they already had a son. Three days after giving birth, the wife returned home with the baby, and I decided to come and offer help," my mother continued. "It was on a Friday when suddenly the father of the baby arrived home. I opened the door for him, saying "*Mazal Tov*, what a beautiful, cute daughter you have. Why don't you throw a *tochom chori* on Shabbat inviting the whole Jewish community?" We then heard his wife screaming from her room, "Hi! Did I give birth to a cat or a dog since you did not visit us in the hospital? Look what a beautiful girl you have! Mrs. Cohen is right that you should throw a party this Shabbat inviting all your friends and she will help you out." He threw a party right after the Shabbat prayers as requested, and a name was given to his daughter there."

"Mom, I have so much respect for you and how you helped everyone, and knew how to organize matters."

"Yes, that's how it was, but don't think it was one-sided. People also helped me since they always remembered the favors I did for them. For example, during my second pregnancy, all my friends helped me and dad spoiled me unlike during the first pregnancy when he was very busy at work.

Exactly on the eve of *Rosh Hashana* of 1941, I easily gave birth to a daughter at home. She was tiny and cute and all of us were very happy. The whole community got ready to prepare a special Shabbat party in honor of the newborn's father, at which she would be named. Unfortunately, this plan fell apart because of a tragedy that occurred to the family of Yosef Simantov. His five-year old son was burnt by a fallen *Yom Kippur* candle. Dad was the first one who tried to save him, but in vain. The child died two weeks later."

Dad was a witness to this event so he may give more details. "We were sitting at home after the fast was over and suddenly heard screams from the synagogue that was close to our house. Immediately, I ran over there and saw the child with flames covering the upper part of his body. I extinguished the flames and took off his burnt shirt and saw that he seemed mostly unharmed. My palms were more burned than his body and it was very painful. The boy went home and a doctor who was summoned determined that he wasn't especially hurt physically but suffered from severe emotional trauma. Unexpectedly, the boy died after two weeks and a cloud of grief befell the family and the entire Jewish community."

"So Mom, what happened with your new baby?"

"We waited through the thirty days of grief for the Simantov family till they came to us and wished us *Mazal Tov*."

"Why did you postpone your celebration," they asked us. "Go and start the *tochom chori* and give a name to your newborn daughter. We really appreciate that you considered our grief and we are willing to *kadamash nik bashad*, take part in your celebration." The next Shabbat after this meeting we indeed made a big celebration, and we named our daughter Esther after Dad's mother. As usual all of our attention was spent on this little child for the first few months."

Dad said, "Back then, I was traveling to many cities in Afghanistan for work. I met businessmen, mayors and diplomats in those places and all of them spoke enthusiastically and excitedly about the first upcoming official competition of the national game of Afghanistan, called *bozkashi*.

On the day of the competition, thousands of people flocked into the capital of Kabul on foot, motorcycles and by cars directly to the huge stadium. It felt like a carnival. All the tickets were sold out, and even tickets offered for ten times the price by scalpers were also sold out. I already had a ticket for the V.I.P. section while my good friend Shimon Shamash didn't succeed in getting a ticket. He really wanted to attend the game and asked me for help. I told him to come to the stadium with me, assuring him that he would get in. He laughed at me in disbelief but decided to join me. When we arrived at the stadium, I met the manager of the team from the city of Kondoz who called me aside for a short talk. He offered me to climb onto one of the team's horses and to enter the stadium along with the rest of the team. I mentioned to him that I was there with a good friend of mine who did not have a ticket and asked if it was possible to let him in. Immediately, he sent one of his aides to take care of Shimon, and both of us were in the stadium.

We had seats a few rows from the King, Zahir Shah's seat. Shimon was amazed that I got him in and thanked me repeatedly.

All of a sudden, we heard trumpets and then the king and his entourage entered the stadium. The huge audience stood up and cheered until the King and his company finally sat. The players were seated on their horses and ready to start the game. The origin of the *bozkashi* game comes from Northern Afghanistan, where most of the population was of Turkish descent. It is a thrilling and fascinating game in which a few teams, each representing a city, compete simultaneously. The horsemen stand in a big circle around a slaughtered goat in the center. When the whistle is blown, each of them tries to snatch the goat, get out of the circle, and run around the perimeter of the stadium with the goat. The others try to block the goat-handler and to snatch the goat away from him by force. Each team receives points for a successful snatch.

The horses chosen for this game are the strongest and the fastest and they are specially trained for this game. The players are the strongest and fastest too. It is a hard and cruel game which requires physical force, a lot of courage, as well as tactics. Players are often injured or even killed. If the goat gets smashed, then it is immediately replaced with a new one so the game can continue until one of the teams win. From time to time, the goat dropped from the player's hands, and another team tried with all their might to pick it up, by bending down from their horses and to gallop away with it at great speed. Another team would try to block and snatch away the goat and move it to the goal. The game was enjoyable and Shimon was very happy that he finally succeeded to see the game that was spoken about for many months.

A week later I returned to the city of Kondoz and met the manager of its team. I gave him as a gift of gratitude a bottle of arak which I knew that he liked as well as a package of high quality pistachios. He was very thankful. One evening, I was sitting with him and a few other guests when he said to me, "*Chojheen-Sahib,* how as a Jewish businessman you are not afraid to carry a lot of money with you? Many people here know this. G-d forbid someone could follow you on your way back and attack you." I instantly answered him, "To the best of my knowledge, only you know that I am carrying a lot of money and no-one else. And besides, as you are my host, it is your responsibility to protect me." He answered, "You are absolutely right."

This is how I went to business in Kondoz many times, with two hundred to three hundred thousand Afghan rupee tied to my back, not concerned at all that someone could rob me on my long trips which lasted a few days. Of course, I was very cautious. I recall one very difficult winter with deep snow and all the roads were totally blocked. I was stuck in a motel that was located among tall mountains at the district of Salang. For almost a whole month there was heavy snow and all the telephone wires were torn down. Back home, my family was worried."

Here, Mom interrupts, "We did not hear from him at all and thought, G-d forbid, that he was gone. Everyone in the community was worried and tried unsuccessfully to contact him by phone. The rabbi of the community, *Mollah* Yosef Aharon, said, "That's it. If he hasn't returned after thirty days, something terrible must have happened."

"What happened, Dad? Where exactly did you get stuck? And what did you eat there over the mountains?"

Dad responds, "There was no permit to cross the mountains and continue to the capital. Every day, I ate just bread and eggs, but fortunately, I am not a big eater and nearby there was an open *nan vahi* and *chi chaneh.* After fifteen days, a car arrived from the opposite direction so I realized that the road was open. I took a lary that traveled cautiously through the snow and after twelve hours, it arrived at the city of Charikar, a one hour drive from my home in Kabul.

Here, we got stuck again, and I met a friend of mine from Kabul named Jaji. He had his jeep and told me that it was forbidden to leave the city until permits were issued. So we remained another day until a permit was given. We moved very cautiously and it took us nine hours to reach home by evening. You can imagine the happiness in our home when a missing person who was thought to be dead appeared healthy and whole.

"What terrible days passed while he was missing! I was heart-broken. Thank G-d he finally returned."

"Mom, you haven't told anything about yourself during those days. How come?"

"At that time, my brother Isachar came to stay with us from Hirat for a month. I'm sure that he has many details to add. When I left him in Hirat, he was very young but then, he was an adult with his own family and a business. From time to time, he used to come to Kabul and Kandahar for businesses. My uncle Jacob and his wife came to stay with us to get medical treatment in Kabul. They were married for a few years but did not succeed to have children.

Hence, I was not alone at home, besides being busy with my little baby girl. I was also nine months pregnant, waiting for the birth any day. Since the fetus was very large, I needed to go to the hospital that time. A physician came to check on me said that a cesarean section was needed. I refused to give my consent to this surgery and asked to consult a professional nurse whom I knew and trusted for a long time. She came and checked me and said that in her view, no surgery was necessary. A day after, she visited me again, and gave me strong massages for about an hour. After this massage, the baby came out naturally; a beautiful, big baby girl.

This was in 1941 when World War II was at its peak. There was hatred towards Jews in Afghanistan since the prime minister was in favor of Germany. About 300 German consultants stayed in Afghanistan at that time and spread propaganda demanding to eliminate the Jews.

Fortunately, the king liked the Jews and did all he could to protect them, like sending police patrols in the Jewish neighborhoods. With

G-d's help, nothing bad was done to the Jews. I stayed in the hospital for three days. On the day of discharge, your father came to take me home. Dad had already organized a *tochom chori* to take place on the upcoming Shabbat. During the party, we named the newborn Bracha, the Hebrew word for blessing, 'May G-d give a blessing of success to our home and to the whole nation of Israel.' A bigger family meant there was more work to be done. My aunt Rivka was hired as a nanny and helped me a lot with the baby. I breastfed her a few times a day, and in-between, Rivka took care of her so I could cook and clean the house.

In 1943 I was pregnant again. It was a fearful year for the Jewish community. The authorities used to unexpectedly visit Jewish homes to look for home-made production of wine and arak. They claimed that Jews sell it to Muslims, which is strictly forbidden according to Islamic law, as Muslims are not allowed to drink alcohol. Almost every Jewish house was invaded by policemen and fifty-two Jews in Kabul, ten of them in our neighborhood, were arrested and handcuffed. The police skipped searching our home since they knew its landlord was the uncle of the king.

The leaders of the Jewish community protested the arrested men. After the intervention of the king, all of them except for two who actually broke the Islamic law, were released. These two were jailed for about two years until the king ordered to release them. The authorities then issued a new law that the Jews could produce wine and liquor only for their own religious ritual purposes and are allowed to be sold only in the Jewish neighborhoods. A special permit was needed to produce alcohol and only two Jews in Kabul received it. At the beginning, all Jews were very scared to break the law, but with time, the authorities neglected to enforce it and many Muslims came to Jewish homes to buy arak, which was a highly demanded product in the capital. Very soon, rumors spread all over Kabul and the authorities discovered that the Jews were selling. Again policemen came to search every home, but this time, they did not arrest anybody, but emptied any barrel of alcohol they found. They did not come by the instruction of high government officials, but by local police officers who enjoyed abusing Jews during World War II to flatter the Germans.

In the year 1944, I gave birth to a third baby girl. This time, I did not need to go to a hospital. We called a midwife who helped me with the birth at home. We named the newborn Tzippora, after my good friend, *Apa* Zippor, who was like a sister to me. We lived together for a decade until she immigrated with her family to the holy land. Of course, I notified her of the birth asking her permission to name the baby after her. She gave her consent right away and said that it was an honor for her. Again, we organized the customary party on Shabbat and the whole community came to celebrate with us. Among them were a few women of Buchara who giggled that we had only daughters. They said to Dad, "What is going on with you, Mr. Cohen? Isn't your wife capable of giving birth to a boy?" Neither Dad nor I responded.

"Dad, how did you feel after having three daughters? Didn't you want a son?"

"In my view, what's the difference between a son or a daughter? After we experienced the tragedy with our firstborn son who passed away after a year, we learned that what is important is that the child is healthy. And besides, after daughters, sons will also come with G-d's help as we believe according to our religion".

Mom interrupted again, "My dear son, we passed different periods in our lives. Happy and sad, but always in a good spirit. Fortunately, we have been very busy people in the family as well in the community, thus, the time passed quickly. I was then busy with raising three little daughters but I still dedicated time to guests.

One day, Dad's stepbrother, *Amo* Morad came to visit us from the city of Hirat. He wanted to try his fortune in the capital like many others. He entered into a business partnership with *Mollah* Yecheskel Kalantar. They received a permit from the authorities to produce alcohol but only for Jewish religious consumption. They were allowed to sell just two bottles a week for each Jewish family but still, it was enough to make a nice profit, which enabled Uncle Mordechai to assist his family in Hirat.

CHAPTER 5

"In 1946 I got pregnant again and this time I prayed to G-d tearfully to give me a son so no-one could laugh at us. I believe that G-d indeed listened to me and I gave birth to a boy. You cannot imagine the happiness in our family. Dad's closest friends sent him a big sheep as a gift, to be slaughtered for the upcoming party that Dad organized. As usual, all of our female friends came to help in this event. First, we made an intimate Shabbat party to honor the newborn's father with just ten very close family members who also brought with them *ashsava*. Then the night before the circumcision, another special party called *Brit Yitzchak* took place. The day after, we made a lot of preparations for the *Brit Avraham*. Dad invited musicians and singers and honored Uncle Morad to be the *Sandak*. A big celebration took place and we named the boy Amnon. Now we felt like an accomplished family with daughters and a son."

"Dad, I assume you felt like you were in the clouds. What more could you ask for?"

"My son, you know me well. I am always cool and I don't get excited, at least not on the outside. I always trusted in the merit of my forefathers and in *Panna Bechoda*. A few years passed and again, many people, mainly from the cities of Hirat and Balkh, flocked into the capital for work. It was the period after World War II and business was at its peak. Many Jews asked for my assistance."

"How? Were you an employment agency?"

"No. Business at that time was conducted as follows: If someone needed to buy merchandise from a big company by credit, good referrals were needed as well as a guarantor. These merchants knew that I was a good place to get this help. I did it many times and thanked G-d that I did not get into trouble from it. Usually, the merchants succeeded to sell the merchandise within three months in Kabul or in another city, then paid back the lender the cost of the merchandise and enjoyed the profits they made. They always thanked me for the trust I placed in them by being the guarantor."

"Dad, what types of merchandise was their businesses made of?"

"Most of it was cloth, like silk, in high quantities which were imported from Japan, India, and the States. The merchants also made business with products which were also exported abroad like *post karakol*, carpets, and dried fruits. Among them were also brokers who dealt in foreign currency exchange and made nice profits. At that time, Kabul knew prosperity and Jewish businesses blossomed."

"Mom said, "The men used to go shopping and the women would prepare various homemade foods. We had no canned or frozen food, but only natural foods like fruit, vegetables, bread, fish, and meat. The fruit and vegetables of Afghanistan are world-renowned for taste and quality and Dad used to buy large quantities and keep them in storage. We always had guests so we needed to be prepared with stocked food.

At that time we had four children, Amnon, the boy, was the youngest. One time, he did not feel good and we got very scared. He was pale and couldn't get out of bed. Physicians were called and medication was given, but nothing helped. The doctors did not succeed to diagnose the sickness and the boy cried and screamed terribly from pain. To our good fortune, a good angel was sent to us from G-d: a righteous and pious man, *Mollah* Shalom Bakshi. One day, he came to us, saw what was going on with the child and said, "All of this occurred because of the evil eyes of people who were jealous of you. He then approached the boy who was lying in bed and started to heal him with mysticism. He took a big, sharp knife and touched the boy's body with it while praying. When he was finished, he asked for an raw egg, threw it with all his strength at a wall, and left. Unbelievably, after a short while, the child's condition improved gradually day after day until he totally recovered.

The daughter of this pious man, Mrs. Chana, was a righteous woman who used to come to us very often to help. She encouraged us during bad times and would help us with every party that we organized during happy times. Once, I said to her, "I feel that one day we will be *Qoda*." She instantly answered, "*Agar saram leiakat Dare*, may we have good fortune to reach it." Our son, Amnon, grew up and indeed married Rachel, the daughter of Mrs. Chana. What a wonder, we had become in-law's after all.

In 1948, the Holy Land was in war for the independence of Israel. Some of the Afghan Muslims were pro-Arabic and again fear had befallen the Jewish community. These Muslims claimed that the Jews did not deserve a state and protested with deep hatred during marches. Fortunately, the king of the country who liked the Jews committed to protect them of any danger and felt indebted to us. One time, the king suffered from an eye illness. All the treatments he received in his country had failed. He was referred to a world-renowned eye expert who lived in Europe.

The king traveled to him, was treated well, and had a successful eye surgery. The king, who did not know that this doctor was a Jew, offered him a lot of money, but the money was refused. The king asked, "So how can I show you gratitude for what you did for me?" The doctor answered, "I only ask you, your highness, to protect the Jewish minority in your country and bestow on them freedom like any other citizen." When the king returned home, he kept his commitment indeed. He instructed the authorities to allow Jews to travel outside of the country whenever they wanted. They needed to fill out a form to get a passport like every other Afghan citizen, and they were always granted.

The state of Israel had been established and until 1952, approximately 3,000 Jews, mostly from the city of Hirat, immigrated to Israel through Iran. The Jewish Agency opened branches in Tehran where the immigrants were first welcomed and assisted in their move to Israel.

Then I asked Mom and Dad, "And what about your immigration?"

Mom answered, "No, we didn't leave then, but my parents, who lived in Hirat, had prepared themselves to move since the economical situation there had declined rapidly. At that time, the Jewish community in Kabul still flourished and needed more religious leaders. One of them who had been just hired, *Mollah* Avraham Yekuthiel of Hirat. He looked for a location to build a new *mikveh* for the Jewish community and found the yard we shared with a few families as the most appropriate place. A river that flowed through the mountains reached the city and a little branch of it came close to our yard. At that place, *Mollah* Avraham Yekuthiel built the pool with measurements that complied with the requirements of the religion and all of us used it for spiritual purification. Even in frosty winters, we dipped into the *mikveh* in order to comply with our religious laws.

In 1948, I was pregnant again. Amnon was already a year old and needed a playmate. Our economical situation was very good; we lived in a big house with a yard and a nanny, Aunt Rivka, who took wonderful care of our children."

"Mom, I am really interested and curious to know how I was born and grew up."

"Pasar-jan, you were born at home on a Wednesday on the eve of the month of Adar (corresponding to February) which is a happy month in our religion in 1949. You were like all of my other children, very similar. We were happy that you were our second son after three daughters and a playmate for Amnon. Again, your father took care of all preparations for the *Brit* and I wanted my dad to come from Hirat to be the *Sandak*. However, it was too difficult for him to come, but at least my brother, Isachar, came from Hirat and we decided to honor him as the *Sandak*. As customary, we made a party on Shabbat to honor Dad and all of his friends were invited. *Ashsava*, which had been already sent by some families as a gift to us, was offered to the guests at this party. On Tuesday evening, the special meal, *Brit Yitzchak*, took place. On the day after, *Mollah* Avraham Yekuthiel performed the *Brit* in the synagogue and we named you Michael."

Then I said, "What a beautiful name. Who chose it?"

"I did," responded Mom. "I always liked this name in particular as I like all names of angels in general."

"And what did Dad say about it?"

"In this regard, he always accepted my decisions in the end. He offered quite a few names which he dreamed about, but they were ridiculous and I did not give my consent.

I turned to Dad, "What do you have to add to all of this?"

"I would just say that I was very happy. What else could I ask of G-d then to have such a beautiful family with five children?"

Mom then said, "After all of these celebrations, your uncle, Isachar, traveled back to his family in Hirat and was welcomed there as a hero."

"What exactly happened, Mom?"

"I suggest you go over and visit him and he will tell you, firsthand and in detail, what happened. I immediately followed my mom's suggestion and took a half-hour trip to Tel Aviv where my 82 year old uncle resided. It was a good break after being together with my parents for a few consecutive days talking and enquiring non-stop regarding the interesting and exciting history of my family.

On my way to Tel Aviv, I passed by the store of my Uncle Nafthali who is two years younger than Uncle Isachar. I parked the car nearby and entered the store to say hello. It was on a Friday morning and Uncle Nafthali was sitting alone. He was surprised and happy to see me. After enquiring each other about the family, I called Isachar asking him if he was ready to see me for a long talk. He happily agreed and I promised that I would be at his home in a short while. Meanwhile, Nafthali and I sat down and I told him that I was writing a history about the family."

He immediately responded, "Yes, your mother told me already, and I have a lot to add. "

"Please tell me about interesting occurrences that Mom may have not told me. So he chose to tell the story of the immigration of his family from the city of Hirat to the land of Israel.

Uncle Nafthali said, "Actually, my parents decided how long to stay in the Diaspora. The state of Israel was established in 1948 and four years later, the state had already absorbed hundreds of thousands of Jews from all over the world and the population grew rapidly. We decided that it was time to try our fortune and to move to our new state. A year earlier, my two sisters, Yael and Shoshanna, along with their husbands immigrated to Israel and my grandparents joined them. Thus, we started to feel a little bit lonely. After thorough preparations, we crossed the border between Afghanistan and Iran in 1952 to the city of Mash'had. We stayed there for three weeks at their Jewish community and then went on to Tehran, where we were welcomed by the Jewish Agency. They accommodated us in a camp opened for new immigrants who were on their way to Israel in a neighborhood called Baheshteya in Tehran. The truth is, we expected to reach Israel very quickly, but we were surprised when we reached the camp. They informed us that we would be delayed there for awhile and had to wait patiently. Later on, we understood why.

At that time, there was a huge immigration of Jews from Iraq and it was too difficult for Israel to absorb so many in such a short while, so the Jewish Agency gave priority to them over the Afghan Jews. The Jews of Iraq were more educated and carried more money with them, so they were put first. We were given medical exams there, and only my mother's results were negative. The Agency's officials said that she could not join us for the trip. We did not know what to do until we came up with an idea to send another woman, a Persian, to take mother's place for the exams.

She introduced herself as Sarah Cohen, took the blood test, and the results were positive. That is how we got the permit to leave the camp. We had stayed there for a total of three months during which my wife was pregnant and gave birth to our firstborn daughter, Miriam. We arranged all of our belongings in suitcases and boxes and waited for the plane to come from Tel Aviv. When it arrived, we all traveled to the airport, checked if any of our belongings or suitcases was missing, and boarded the plane. The plane took a longer route, over Turkey, since it could not fly over hostile Arab countries on the way to Israel.

We arrived in Tel Aviv and then realized that only one box from our belongings was missing. I would say it was the most important box in our possession. It contained ancient, holy books that belonged to grandpa and he valued them all his life more than anything else. We searched for the box all over the airport but it was totally gone. After a few days, the airport personnel found a few scattered books and delivered them to us.

At the airport, we passed through registration procedures and a house was assigned to us, located in an isolated village called Beth Shemesh. A minibus took us to that place on a very rainy night.

When we arrived at our destination, grandma was stunned by what she saw. "Where are we?" She asked. "I only wanted to go to Jerusalem. This is why we came to Israel. Here, I see only a desert and I am not getting off the bus." We tried to calm her down but it did not help. After a lot of begging, she agreed to get off the bus, but then another problem arose. The ground was covered in mud and we were stuck in it and could hardly move. Uncle Isachar had new shoes which were completely ruined. Finally, we reached our new home and immediately fell asleep from exhaustion. When we woke up in the morning, we looked out the window and we saw a desert with mountains and rocks. Afterwards, we realized that even food would be hard to find.

The men were required to work with huge axes they received to break rocks from the mountains for construction purposes. After two weeks, it was getting close to *Pesach* and we decided to leave for Tel Aviv to our sisters, who came to Israel before us. Israel was a very poor country and food was rationed for each family. Each family was entitled to buy only two hundred grams of meat per week and we had to wait in a long line to get it. Chicken and sugar were not available at all and we had to walk long distances to get water. We suffered severely and grandma was crying all the time. She complained, "Why did we come here in the first place? What was wrong with Hirat? Is this the promised land?"

He continued his story. "We did not return to Beth Shemesh again and we gave up the house we received from the Jewish Agency.

For *Pesach*, we stayed in Tel Aviv with our sisters and then we bought a subsidized house with our money and loans we received from our friends of Hirat. This house was in a neighborhood of Tel Aviv where most of the inhabitants there were new immigrants from Afghanistan. The first year in Israel we experienced unpleasant adventures; we endured only troubles and difficulties. The worst was our bad feelings during *Pesach* and *Sukkos*. We missed the giant *Sukkah* we used to build in Afghanistan and the many families we invited to join us.

Here in Israel, little by little we made advances. We worked very hard in order to save money and to make our lives more convenient until I reached the feeling that there is no better place for Jews to live than in the land of Israel. This is our land and in my view, all the Jews have to return to it. Now the country is well-developed and the quality of life is among the top countries in the world. I visited many countries in Europe and I have not found a nicer and safer place to live than in Israel. I know you would ask about the terror we suffered and I can assure you that nevertheless, it is still more safe to live here than abroad. Nowadays, all the countries in the world live under fear of terror and Israel is no exception."

I turned to Uncle Nafthali, and told him that our time was over and my other uncle was waiting for me to continue this type of discussion.

"Yes," said Nafthali. "I am sure that Isachar has a lot to tell you about our family. In his youth, he was a very brave and courageous man." We embraced and kissed each other and I left.

After a ten minute drive, I reached the house of Isachar. I knocked on the door and his daughter, Osnat, opened it. "Welcome! Please come in. Dad is expecting you," she said. Immediately, I saw Uncle Isachar who looked well and he invited me to sit down and talk. Usually, he does not feel well since he had an open-heart surgery a few years ago but now, I found him in a good mood. We sat down on the sofa with me in the middle and my uncle and his daughter at my sides. While they started asking me about life in the states, my uncle's wife came in with fruit and drinks to welcome me. She wanted to bring more food and I had to stop her saying that today was Friday, and she had to prepare for the Shabbat. Then I started the real discussion with my uncle.

"I would like to hear from you first-hand why people that know you describe you as a hero."

He responds, "Okay, I will tell you briefly. Once, a month before the holiday of *Shavuot*, a group of Jews and I were at our *baq*. It was spring. Flowers were blooming all around and it was pleasant to sit outdoors. Whole families, including women and children were there relaxing.

Suddenly, an evil man who had a crazy hatred of the Jews, appeared out of nowhere with a sharp knife tied to his hand. He immediately attacked *Mollah* Meir Amballu who was holding a little child in his hands. His wife, Rachel jumped at the attacker and moved his hand away from her husband but the knife caused a deep cut in the child's face. I saw what happened from a distance, so I ran quickly towards the attacker and with all my strength; I pushed him into a nearby canal. Other people started to gather around and I jumped into the canal in order to remove the knife from him. I did not succeed since it was knotted to his hand. I held his hand tightly and asked people for a knife to release the knife from his hand. Someone gave me a pocket knife and I cut off the attacker's knife. We took him out from the canal and tied him to a tree until the police came.

"How old were you at the time?" I asked.

"Twenty-three," he answered.

"And weren't you afraid at all?"

"No, in such moments you don't even think about fear but only about how to protect yourself and innocent people. In a short while, policemen arrived, beat him terribly, and took him to jail. To our surprise, the judge sentenced him to only one month in prison. A month later, we returned to our homes to prepare for *Shavuot*. The eve of the holiday came on a Friday night.

I went up to the second floor of the synagogue while many women were sitting at its yard. Suddenly, I noticed the same attacker entering the yard. Immediately, I ran down to catch him. Unfortunately, he already succeeded to stab five women. I struggled with him and another friend, Abba Cohen, came to help me. Abba Cohen got stabbed in the back. Afterwards, the attacker attempted to run away but policemen, who arrived at the scene after hearing screams, succeeded to stop him. Most of the congregation, including my parents, left for their homes after an interrupted service and locked their doors. I remained to aid the injured. I approached the *Qomandon* whom I knew personally, who likes Jews and was always ready to help us in case of trouble. He immediately called a physician and a few nurses of a nearby hospital to take care of the injured. From a far distance, I heard my mother scream, "Isachar! Enough! Come home immediately!" And I answered, "I cannot leave the injured in this situation. Someone of the community has to control the scene, so I will!." The physician examined my friend, Abba Cohen, and diagnosed him as severely injured.

The knife had penetrated deep into his back and there was a high chance of infection. The physician prescribed him antibiotics and I ran to a pharmacy that was open that night on duty to pick it up. Meanwhile the physician and the nurses took care of the five injured women of whom, two had died at the spot. When I arrived at the pharmacy, I told the pharmacist that someone was injured and we needed the prescription to be prepared urgently. I also told him that I was Jewish and this night was Saturday and a holy day for us so we do not carry money on this day but I promised to pay right after the holiday. The pharmacist absolutely refused even though I tried many ways to change his decision. I returned to the scene and notified the *Qomandon* what happened. He immediately ordered a policeman to go to the pharmacy and to bring either the medication or the pharmacist with him. To our surprise, the policeman returned empty-handed with the pharmacist. The *Qomandon* asked the pharmacist why he did not have the medication.

The pharmacist answered, "Only when I receive money do I give out medication." The *Qomandon* slapped him hard and said, "You will bring the medication immediately or you will be arrested." The pharmacist understood this language; the medication was brought right away and my friend was saved from infection and its possible complications. Regarding the injured women, it was a real tragedy since all of them belonged to the Ovadia family. This holiday was terribly interrupted and turned from joy to grief."

I asked him, "What happened to the attacker this time?"

"The *Qomandon* decided not to take any risks and sent the attacker handcuffed in a police vehicle to the capital Kabul to be sentenced there for his crime. An affidavit signed by the *Qomandon* with a full report of what had happened was submitted to the court and the attacker was sentenced to be hanged."

I asked my uncle again, "How come he was released so fast after the first time he was arrested?"

Isachar answered, "We asked this same question in the community many times and the answer we got was that the attacker was a person with a weak character who was incited to attack Jews as revenge by his brother. The two brothers belonged to a gang of fanatic Muslims whose hatred for Jews was without limits. It was the period right after the Israeli War of Independence which established the Jewish state. To our good fortune, the leaders of the Afghan people liked the Jews of their country and prevented any hate crimes."

At that point, my uncle's wife asked me to stay for lunch. I respectfully declined the nice offer, saying that today was Friday and I promised to have lunch with my parents.

Isachar then continued, "I have one more interesting story that I would like to share with you before you leave. I was a child, ten or eleven years old, when Grandpa Mordechai, who was a renowned rabbi and preacher in the community, came to his last day of life. It was on a Friday evening before Shabbat. He called for all his grandchildren to be around him during the *Kiddush*. All of them gathered around his bed and started to sing Shabbat songs which they learned in the *Yeshiva*. The song, which they sang during his last minutes, was an exciting one, which we used to sing on each important event in the family and sounds as follows:

Aya aya aqa, chalas bakon mara

(Come, our Father, and provides us release)
Azain azain galot, nejat farma mara
(Thus, we will be saved from being displaced)
Bia bia chaliq, azma razi beshoo
(Come, our Creator, and be pleased from us)
Bia bia padsha, mara rahbar beshoo
(Come, our great King, to lead us)
Gona gona kardim peshaiman shodim
(For our sins, we express regret)
Galot galot oftadim, tabah shodim
(We fell into exile, and need to repent)

During our singing, he returned his soul to G-d. This was a spectacle that I would not forget forever. Right after, the Eve of Shabbat began and my dad instructed all of us not to ruin the holiness of the Shabbat and not to cry as it is forbidden on this holy day. We prayed as usual, then ate the Shabbat Eve meal and in the morning, went to the synagogue as usual. We returned home for the Shabbat meal and then tried to rest as we used to. But who could rest in such circumstances?

In the evening, when the Shabbat passed, we made *Havdalah* and right after, the screams of grief began. Dad, who was very close to his father, took it very hard and was devastated. The next day, on Sunday, grandpa was buried in Hirat in a huge funeral with great respect and eulogies."

With this, Uncle Isachar finished his stories and invited me for lunch. I politely declined again as it was close to Shabbat and my parents were waiting for me. We embraced and I left. I drove quickly back to my parents and as I arrived, Mom was already standing at the front door. "Where have you been?" she asked me. "Why did it take you so long? Don't you know that today we eat lunch on time?" I told my parents about my exciting visit by Mom's brothers, Nafthali and Isachar, and their stories and my parents verified them as true.

CHAPTER 6

Mom, Dad and I sat down for lunch, and after we finished and had a short rest, we sat again to continue with our stories exactly at the point where we stopped – the period when I was born.

Mom began, "You were a quiet and disciplined boy and we hardly encountered any difficulties in raising you. I cannot even recall any specific problems. You were a chubby baby and our neighbors liked to play with you. The rabbi of the congregation, *Mollah* Yosef Aaron, who did not have children, especially liked to play with you. He used to hold you in his arms and to call you *Sosorai*."

Since I never heard this word before I asked, "What does sosorai mean?"

Mom answered in one word. "Chubby."

"All of my children were raised with much love and I breastfed each of you for a very long time. All of you grew up in our house like children of rich families. You personally, had a lot of luck as you had older sisters who always protected you. And do not forget your nanny, Aunt Rivka, who helped to raise all of my children with much devotion.

"Dad, as usual, was busy traveling or with providing the needs of the Jewish community which grew steadily and a lot of public work had to be done. As I told you already, Jews from other cities joined our community and we always provided them with help."

Dad said, "I recall a Jew named Daniel who arrived from Hirat in a terrible condition. He was a crazy man and everyone called him

Daniel Divaneh. He used to sit at the front of stores that belonged to Jews, dressed with dirty and torn clothes and did nothing. He was very dirty, unshaven, and had long hair and dirty, long nails. Naturally, no one wanted to get close to him.

The Eve of *Pesach* arrived and on that morning, while the congregation was in the middle of the prayer, he rushed into the synagogue, and with a threatening voice, shouted, "Are you Jews? How come you don't have any compassion for me? This night is *Leil Ha'Seder* and I have no place to go to and celebrate like every Jew should. I am really considering converting to Islam and then this sin will be on you. Is this what you want?"

Complete silence filled the synagogue but no one volunteered to host him. I stood up and said, "Daniel, follow me. You are my guest as long as you want to stay with me." We left the synagogue together and I gave him money so he could go to the barber and afterwards to the *hamam*. An hour later, he appeared at our house. I gave him a clean shirt and pants and could not believe it was him, a completely new person. His mood and behavior were totally changed for the better.

Towards evening, the house was clean and shiny and the dinner table was prepared for the *Pesach* meal. I said to him, "Daniel, come with me. We are going to the synagogue." Arriving there, everyone of the congregation was surprised, asking me, "Who is this guest that you have with you?" I said that he was a good friend of mine. Then the rabbi came in and the prayers started immediately. At the end of the service, everyone wished each other a happy holiday and then suddenly, someone of the crowd identified my friend as Daniel Divaneh. Rumors spread all over the crowd and they could not believe. For them it seemed to be a real miracle.

A few days after being with us, he started to play with our children and looked after them. He followed instructions and I left him in charge of cleaning the yard and buying groceries. Our family gave him good feelings and enhanced his self-esteem. He turned out to be a good assistant and stayed with us for a few years until the establishment of the state of Israel. He then left to be among the first new immigrants to the newborn Jewish state."

Then Mom said, "I must admit that Dad, though not being a big speaker, was a big doer. A very active, initiative person. That is your father that I have known since we were married."

Dad then said, "I do not know if that is how I am, but maybe Mom knows me better than I know myself. Anyhow, I think that she always used to overestimate a bit."

I noticed that Dad was getting tired or maybe he just did not want to talk about himself personally in a way that would sound like bragging. Hence, I encouraged him to continue telling me about general events that occurred in the community at that time period (1948-49) that did not involve him. He immediately accepted my offer and started to tell about a tragedy that befell the community unexpectedly.

One day, two youths, Mayer, the son of *Mollah* Haji Yitzchak Basal, and the other named Benjamin, the son of *Mollah* Reuven Amram, totally disappeared. They were at the beginning of their lives and were dealing with foreign exchange currency. It seemed that because they were young and inexperienced in life, and were not cautious enough in this type of business. Hours passed and there were no signs of them, so the police were called into action. After thorough searches and interrogations, the police questioned the last people that saw them. Suspicions fell on a maintenance man in the American embassy who supplemented his income by dealing in currency exchange. He admitted that he met them a short while ago and sold them American dollars but they left him right after. He submitted a reliable alibi, which was verified by the police and they left him alone.

At that time, the fears reached to a climax and the whole community, adults and children alike, joined the police in the search. At the end, after another several hours of searches, the police got back to interrogate other two maintenance workers at the American embassy. They remained the prime suspects and the police wanted to search their private homes. Unfortunately, it was impossible for the necessary search to be done because these two employees lived in a building which was part of the embassy's complex, under the protection of diplomatic immunity. The only way to get in was for the concerned family to ask the American ambassador for a search permit on a moral basis. The ambassador immediately understood the sensitive situation and granted the permit.

Haji Yitzchak, the father of one of the missing boys and a very smart man, hired a few experienced laborers and together they went into the residential section of the embassy complex to make a very thorough search. They moved from room to room and from yard to yard and suddenly, Haji Yitzchak noticed that one of the store rooms looked different from the others they had already searched. Each store room was built with no flooring right on the sand and in this specific one, the sand was visibly wet. Even though most of the space of the store room was covered with luggage, this wet sand triggered his suspicions. He instructed the laborers to move out all the boxes and luggage and to start digging into the sand. It was already the late hours of Friday, so he urged them to work very fast and intensively in order to finish before the Eve of Shabbat. Even at that sad time, he was concerned about keeping the Shabbat.

Suddenly, during the digging, a horrible sight appeared before them. An arm poked through the sand. The laborers then continued to dig deeper at that spot and two corpses were recovered one after the other. The community in general and the family in particular, could not have expected such a horrible tragedy. Deep grief and mourning befell the whole community. Such an event was very rare; everything comes from G-d. Their time had arrived.

My Mom then said, "We are now in the early 1950's and the family expanded. I gave birth to a daughter after two sons. As usual, we made all the regular rituals concerning this birth which included giving her a name. Friends said to us, "You have among your children a boy named Amnon, so it is appropriate to name the girl Tamar as King David had a son and a daughter named Amnon and Tamar." We liked this idea and so named her Tamar. You were two years old and your older brother, Amnon, was a little more than three years old. You boys were good children and grew up without causing us any difficulties. Your older sisters and nanny, aunt Rivka, helped and looked after you. Thank G-d our economical situation and security were good and we had nothing special to worry about besides having more children. We had a big house surrounded by a huge, fenced *havli* where you children played in all of your free time. We had in abundance all of the necessities and we lived a quiet life embraced by a supportive community.

At that time, we received news from Hirat that Grandpa Yeshaiau, the father of your Dad, passed away at an old age. Dad was very saddened at this news but he accepted it as a fact of life. Soon after, a new baby boy joined our family and Dad named him Yeshaiau, after his father. The birth occurred very close to *Pesach* so the nearest Saturday in which the newborn's father is honored was on *Shabbat Hagadol.* The celebrations concerning this birth were multiplied because of the holiday and many people took part in them. The peak was the *Brit Milah* celebration accompanied with many appropriate songs."

"*Pasar Jan*, from now on I guess that you yourself remember what went on in our family and community."

Dad suddenly raised his voice, "Listen! You already know where we lived at the beginning, at *Shahar Kohna*, and then we moved to *Shahar Naw*. In the 1950's our community was growing fast since many families migrated to the capital from other Afghan cities, especially from Hirat, Balkh and also from Buchara of South Russia. Do not forget to mention that the synagogue was adjacent to our house for more than ten years."

"Yes Dad, I remember everything from that time but I would rather hear it from you as your description is more accurate and detailed."

Dad got a bit uncomfortable and said, "I am sure that you remember well all the events that we passed through since you lived with us, so I suggest that at this point we have to stop and you will continue the story of our family by yourself."

I understood that I bothered my parents too much, but I felt that I would not have another opportunity in the near future to get such precious information out of them. It was the last days of my stay in Israel after an exciting three weeks of being with my parents and hearing from them, first-hand, the history of their way of life, how they passed through good times and bad times.

Very soon I was going to separate from them and go back to New York. The upcoming separation caused me a bit of sadness, but on the other hand, I knew that I was going to leave with an excellent feeling of satisfaction and achievement of all of what I experienced during this visit to Israel. I was amazed to realize that in such a very old age, my parents had the patience to sit down with me, to tell and explain to me all of their past.

I felt that I could not leave before asking one or two very personal questions. It was not easy for me to ask such questions so I thought of how to do it in the best way. "Dad, Mom, I have a very personal question for you. What is your secret to reach such an old age and also be respected?" There was a moment of silence and my parents looked at each other with shy smiles.

"What does he want from us?" asked Mom. She then said, "Listen *Pasar Jan*, in this regards there are no secrets. Everything comes from G-d. A person determines his own way of life like a gardener who takes care of a garden. If he treats it well, by planting trees and flowers and watering them, he receives positive results: fruitful trees and beautiful flowers. If not, he receives thorns and weeds. Exactly like that is the relationship between a human being and his friend. If you do favors for the other, you get back favors, sooner or later and vice-versa. Dad and I have been together all the time and we look at life optimistically."

"We received this trait from our forefathers and because of their kindness and in their merits, we reached where we are now. You should know that each favor you do is recorded in Heaven as we Jews used to say "The notebook of G-d is open and his hand writes in it." We are hopeful that you, our children, will continue this pattern and keep to it strongly. I know that Dad would agree with me. Here, your father sits, listens, and nods. Each day that passes and he is in good health is because of the good deeds he performed and the respect he gave to his parents and parents-in-law, which is known as a remedy for a long life. Whenever Dad was physically close or far away, he always honored and respected his parents and mine. On every occasion, he used to send them presents, delicious food, and more. He never hurt any human being or harbored hatred against anyone. Even though he was, during those years, a public figure in the community, he did not have any enemies.

I am aware that that generation was a different one than today. I do not know how to describe the generation of today – a fickle and foolish generation in which people forget their origins. It is a shame that some Jews forget their identity and that our nation is different from other nations. Our land of Israel is a holy one and living there does not automatically make you a good person if you stick to dirty ways. We inherited this land from our forefathers on the condition that we follow the Holy Torah, but if Jews do not comply with the demands of the Torah, we are in trouble."

Suddenly, my Dad sat next to me and wanted to say something. I said, "Okay Dad, as we come close to the end of our interview, maybe you can say a few words of blessing."

"Yes son, but before blessing, I would like to add something for you: Keep away from anger since all diseases come from it. Regarding all the rest that Mom mentioned, I totally agree with her. Our heartfelt blessings to you and we wish you much success. Take care of yourself well and always remember us. We wish you that in your next visit, you will come to the Holy Land to visit us with all your family members."

The next day, we separated with hugs and kisses. Mom, as usual, followed me down the staircase and gave me a bottle of water as a charm for success, as well as a bag of fresh fruits of the Holy Land.

She whispered, "Will we have another opportunity to see you again?" I traveled with the car I rented during my stay in Israel towards Ben Gurion Airport with very good feelings of satisfaction as if I revealed a whole new world. My parents liked the idea of writing a book about the history of our family and of the Afghan Jewish community. They instructed me over and over not to stop this mission that I took upon myself until I bring it to an end. Now I am continuing to write this story of Dad's home in Afghanistan by myself, from the earliest age from which I can remember, the 1950's and 60's.

Of course, I am combining with my own memories' stories with those that I heard from my parents, and I shall place them in the proper sequence of time.

CHAPTER 7

After two months of writing at my home in New York, researching and interviewing friends and relatives to find more stories, I decided to return to my parents to check with them, firsthand, many details in order to get a more comprehensive picture. I was in Israel again and right away, I continued questioning my parents.

I turned to my parents and said "Let's discuss the way we handled our Jewish customs in a foreign country, as we did in Kabul. I will tell you all of what I remember since the earliest age of four years old. Listen to me carefully and just respond if it was true or not."

"Okay," said Mom and Dad, who nodded in agreement that it was a good idea.

"Dad, in my previous visit, you tried to describe the beautiful seasons in Afghanistan. Each season had its correlated Jewish holidays. The Hebrew months of *Adar-Nissan* (March-April) come in spring. At that time, we would prepare for the upcoming *Pesach*. Each house had to be totally clean of *Hametz* by the arrival of the holiday. The women of the community were in charge of this difficult job and they would come every day to other houses to do that cleaning voluntarily for no pay. All foods eaten during this holiday should be strictly Kosher for *Pesach*, which means that the food should not contain even a crumb of *Hametz.* Thus, even ingredients, like sugar, salt, spices, and dried fruits, were produced within the community under rabbinical supervision. Regarding oil kosher for *Pesach*, we had no problem since at that time, sheep fat was available. The most complicated task was baking matzos.

The minute I mentioned the word, matzos, I noticed Dad nodding and sitting straight as if he was about to say something.

"Yes, Dad?" I said to him.

"I was in charge of baking the matzos in the community. The whole process of preparing the matzos from its beginning till the end lasted approximately two months. We rented a flour mill at a nearby village. A few youths cleaned the mill thoroughly so absolutely no *Hametz* would remain and they changed the big stone that grinds the wheat into flour. I was busy every day of the three weeks before *Pesach* in baking matzos. Each day, I baked for two to three different families who supplied the flour themselves. I did not have any time left to rest and of course, could not go to my own business. Actually, I was the only one in the community who could do this baking properly since no one else had as much experience as me in this task.

One interesting episode that I recall from that time is that one early morning, right after the daily prayer; I had to go to my own job. On my way, I saw *Mollah* Babajan Vardi with his wife and a sack of flour in his arms. They greeted me and asked for help to bake their matzos. I instantly answered that it was impossible today as I was urgently called to my own job and I told them to bake it themselves. He answered, "No, Mr. Cohen. You must help us and I will take care of your private job. Just tell me what has to be done." I answered him, "No, you cannot help and it is very urgent." "What is it about? Maybe I could help." "I urgently need to raise 100,000 Afghan *rupees* and to make a *hvale* to the city of Kandahar. A Muslim there has been expecting this money for a few days and it would be really unpleasant to delay any further."

I could not explain to him that I did not make the transfer until now because I was very busy baking matzos for *Pesach*. He would not understand. On the spot, *Mollah* Vardi responded, "I have this amount in my office and I could make the transfer for you right away and you will pay me back when you can. Take the flour from me and my wife will stay with you to help you bake the matzos."

Of course, I could not refuse him anymore and the deal was done. I enthusiastically baked the matzos for his family, and he made the money transfer for me on the same day. Usually, I baked the matzos for every family in the community and did not work at my job during that whole period until the eve of *Pesach.*"

Mom said, "Others made money and businesses and your Dad was busy handling the necessities of the community."

Suddenly Dad interrupts," What did you say? I do not even think of what others have done. I knew what I had to do and I would not leave my responsibilities to the community. As the holiday was approaching, more and more work has to be done to prepare for it."

I was a little child and I remember well all of what had been done in our home during those days. We had special utensils for *Pesach* which had been put aside during the whole year. Nevertheless, we needed some more utensils for *Pesach* so we had to make *hag'alat keilim* of the ones we used during the year. At that time, disposable utensils were not available.

Dad and Mom laugh, "Disposable? We did not know what that was."

Then I remember how the houses were cleaned thoroughly. This is really a story by itself. For example, a very strong man would come to each home, collect the huge carpets, take them to the nearby park, and put them on the lawn. He would shake the dust out of them since there were no vacuum cleaners like today. He would scrub the carpets by hand and wash them. After he was finished with his job, he would return the carpets to the homes. He did the same thing for the carpets of the synagogues. This was a big job until the last minute of the eve of the holiday. Searching for the *Hametz* in each house and its yard took many hours to be done.

Dad nodded his head at this point and said," This is true. We have been very strict in searching for it." Finally, the Eve of *Pesach* arrived. All of us dressed up nicely, with new clothes that were specially made by a tailor. Dad nodded again and said, "I remember it very well. To choose the garment and to go to the tailor was the peak of happiness for the children." At that time, there were no boutiques and tailors did all the quality clothing. Dad took all of his children to a shoe store to pick up new shoes special for the holiday. We always used to buy shoes that were a half size bigger. Dressed beautifully, the whole family went to the synagogue. All the people there were dressed nicely and there was an atmosphere of celebration.

After the Eve of *Pesach* prayer was over, we went home to find a long, prepared table with all the special food and wine on it. We loudly read the *Haggadah* and sang the songs that were in it. The next night, this same ritual repeated itself again into the late hours. During the days of *Pesach,* we used to make *Moed Bini* for our friends and acquaintances. I remember that it was a very nice custom to visit each house starting from *Shahar Naw* to *Shahar Kohna* by foot. Also, our own house received many visitors during the holiday, including non-Jews, who came to wish us an *Ed Fatir*.

Then Mom continues, "I will never forget the last night of *Pesach Shab-sal,* with the party that I prepared to honor it. In Israel, this celebration is called *Mimona,* but it cannot be compared to the great parties we used to make in Afghanistan. Our table was covered with many types of sweets including the beloved *nishala*. As soon as the holiday was over, the children used to run to the fields and bring home *sabze gandom,* which was considered a symbol of blessing and success. Every head of household used to bless his children on this day, *Isru Chag,* and in the synagogues, the rabbis would bless the whole congregation, especially the children, with success.

We baked bread after many days when we could not eat any bread, and we went to parks to eat it with barbecued meat. This was a custom each year and it ended with the wish "Next year in Jerusalem."

Mom said, "These days were such great ones that you could not even dream about them. It was the end of spring, towards the beginning of the summer and its mild and pleasant warmth. We used to go to our summer homes on the mountains where the climate was cooler and took with us supplies of lamb, meat, and vegetables. Some of the families stayed there for the whole summer which lasted two to three months. This was a different world, something like paradise. It was a joyous time for the children as well as for us. During this period of summer, we commemorated three weeks of mourning for the loss of Jerusalem in ancient times, during which, we did not eat meat. On the last evening of these weeks, the ninth of Av in the Jewish calendar, the date of the destruction of the temple, we placed carpets on the ground outside and the whole congregation sat on them and prayed with a sad melody during fasting. This ritual was repeated the next whole day as well. Everyone was fasting, including children, women, and the elderly. Even the sick people fasted.

Dad said, "I especially remember *Mollah* Shimon Gol, who used to cry bitterly during these mourning prayers. He seemed to fully identify with the destruction of the temple in Jerusalem which occurred approximately 2000 years ago and we Jews have not merited rebuilding it. At this point, Mom and Dad simultaneously declared that it is not easy to be a Jew in this world."

Mom adds, "At least now we have the land of Israel and all Jews should come live here instead of remaining in the Diaspora."

"What?" said my father. "Here, we are also in exile! We do not have, in this secular state, enough freedom to practice the religion and the land of Israel is far from peace and tranquility. May G-d help and have pity on us and on our land."

"Let us continue with the beautiful seasons of Afghanistan and their correlation to our holidays." I said. "The end of the summer arrives and we start with the special prayers of repentance which last a whole month until *Rosh Hashana and Kippur*. Every day during this period, we started prayers very early in the morning, around four a.m. I remember well Rabbi *Mollah* Yosef Aaron, who lived in our *havli* and used to practice blowing the *Shofar* during these weeks before the high holidays. On the two days of *Rosh Hashana*, he was the one who blew the *Shofar* for the congregation. He used to cry while telling us the renowned story of Rabbi Amnon in the Afghani language and the whole congregation joined him in his sobs. He prayed loudly, with an Afghan melody that was heartrending, from the prayer book, "On the first day of the year it is inscribed, and on the Day of Atonement the decree it is sealed, how many shall pass away and how many shall be born, who shall live and who shall die, who at the measure of man's days and who before it; who shall perish by fire and who by water, who by the sword, who by wild beasts, who by hunger and who by thirst; who by earthquake and who by plague, who by strangling and who by stoning; who shall have rest and who shall go wandering, who will be tranquil and who shall be harassed, who shall be at ease and who shall be afflicted; who shall become poor and who shall wax rich; who shall be brought low and who shall be upraised. But Penitence, Prayer and Charity avert the severe decree."

A week after *Rosh Hashana* comes *Yom Kippur*. As customary in the Jewish religion, the holiday starts on the eve before and lasts for about twenty-five hours. At the eve of *Yom Kippur*, the synagogue was crowded with the whole congregation, including women and children. Everyone was dressed in white, formal clothing, and in my view, they looked like angels. The start of the evening prayer is with the famous song, *Kol Nidrei*, sung loudly by one of the members. The position of the singer was sold by auction and almost every year, *Mollah* Shalom Shamash would win this auction and pay any price."

Mom nodded her head in agreement and said, "We, the women, heard in a partitioned room, the prayer of *Kol Nidrei* in a low voice and we knew that it was *Mollah* Shalom Shamash. After many years when he left the congregation, *Mollah* Eliyaju Amballu replaced him and also paid any price.

When Dad heard the name of *Mollah* Shalom Shamash he immediately started sharing information, "He was one of the top leaders of the Jewish community from 1950 to 1960. He was a very prominent person and it was a pleasure to be with him. We had a friendly relationship and he really appreciated what I did for the community. I always had a few opponents who did not agree with what I did, but *Mollah* Shalom always encouraged me and told me to pay no attention to them. Once, he told me that he used to buy the right to sing *Kol Nidrei* since his birthday was on *Yom Kippur*. His origin was from Mash'had in Iran. His grandfather, *Mollah* Avraham, would risk his life to turn his store into a synagogue on Jewish holidays in a time when it was strictly forbidden by the Iranian authorities for Jews to gather together in prayer."

"Yes, Dad. I heard about this family a lot from the son of *Mollah* Shalom, Yaakov, in New York. Please correct me if I make any mistakes. From the age of seventeen, Shalom was already a public figure. It has been told that he saved many Jews at the time of their expulsion from the city of Mazarsharif. Another story about him that is told is about a Jew who traveled by train near the Russian border and changed cars during his travel. Unfortunately, he fell off the train and was smashed to pieces. *Mollah* Shalom appeared at the scene, patiently collected all the pieces of his body, and laid them to rest as if he belonged to *Chevra Kadisha*. During the day of *Yom Kippur*, the whole congregation stayed at the synagogue in fasting and prayer until it was evening. Then they speedily went home to eat the first meal after fasting all day.

I remember that right after the meal, we used to leave home and started building the *Sukkah* for the upcoming *Sukkos* holiday in a few days. This holiday commemorates the stay of the Jewish people in huts during their wandering in the Sinai desert on their way from Egypt to the land of Israel. It is celebrated for seven days while each family eats all of their meals in the *Sukkah*. The very religious would even sleep in it. The roof of the *Sukkah* must be, as in ancient times, of only leaves or wooden branches. We used to build a big *Sukkah* to accommodate four families and covered the ground with huge carpets that Dad specially brought from friends who had a carpet business. The women were in charge of decorating the *Sukkah* from the inside.

On this holiday, we pray in the morning while holding in the hand three types of plants together with an *Ethrog*. All of these four symbolize the four types of Jews that exist. One of the plants is the *Lulav*. The other two are the *Hadas* and the *Arava*. All of these four grouped together are called "*The Four Species*". The *Arava* and the *Hadas* can be found anywhere in Afghanistan. We picked them up from a park near our house. The other two were more difficult to obtain. A few of my friends traveled to the city of Jalalabad near the border of Pakistan to find the *Lulav*, and the *Ethrog* was ordered from the Holy Land.

Dad immediately interrupts, "I recall that one year, no one of the congregation succeeded to obtain an *Ethrog* for the upcoming *Sukkos*. Miraculously, at the last day before the holiday started, our family received, via airmail, an *Ethrog* which was sent from the land of Israel by Grandpa Yehuda. Of course we shared this *Ethrog* with the whole congregation. During the seven days of the holiday, our whole family, including three others, were dressed formally and ate meals in the *Sukkah*. Some of us even slept there overnight. The whole city was aware of this Jewish holiday and even Muslims were invited to the *Sukkah* and wished the Jewish community, *Idbid Bandan*. Actually, this holiday was a great celebration for the Muslims. The grocery stores, fruit and vegetable stores, and fish stores made plenty of money from us. During the seven days of *Sukkos*, we did not work out of respect for the holiday.

Right after these seven days, we celebrate one more holiday, *Simchat Torah*. On this day, we danced and sang in the synagogue with the Torah to show our love and devotion to what is written in it. This day is the end of a sequence of consecutive holidays which started over three weeks before with *Rosh Hashana*. On this day, it was customary in our community to visit friends, in particular the prominent ones. Many people of the community from all classes visited us on this day just like on *Sukkos* since they respected us. Mom prepared all kinds of delicious food and offered it to the guests."

I turn to Dad and inquire, "Because of your close relations with many Muslims, how would you describe their feelings towards the Jewish community?"

He answers, "My absolute impression was that they liked us. We had total freedom in the country and never felt organized hatred against us. Of course, here and there, there were some expressions of hate but it was on a personal basis and very rare. I would like to bring to your attention that many Afghan tribes, especially of the south, originated from ancient Jewish tribes who disappeared over time. Their faces looked different from the original Muslims. Their skin and hair were lighter and they still kept Jewish customs despite losing their Jewish identity. Even though they have been out of the Jewish religion for thousands of years, their wives used to light Shabbat candles and went monthly to the *mikveh*. They ate only kosher food and also had their own separated cemeteries. It is very interesting that their tribes had names similar to the ancient Jewish tribes, such as *Rabani*, parallel to the Jewish tribe of Reuben, *Shinwari*, parallel to the Jewish tribe of Shimon, or *Lalwani,* parallel to the Jewish tribe of Levi, *Jaji* as Gad, *Ashuri* as Asher,*Yosuf-za* as B'nei Yosef, and *Afridi* parallel to the Jewish tribe of Efraim, and so on. These *Pathan* people, who extraordinarily liked the Jews, traveled far distances to make business with us."

Mom breaks her silence and says, "Why do you go so far? Let's talk about our local neighbors. Opposite from us lived Mohammad Zahi, a cousin of the king, who was the governor of the city of Pole-Chomry. His family used to visit us often. His little daughter, Raziya, came to our home almost every day and used to play with you when you were a little boy. One day, while we were sitting in our yard, we suddenly heard a knock on our door. It was the prime minister of Afghanistan, accompanied by the king's first officer, Sar Yaver, who was an old friend of Dad. Dad knew the prime minister too from previous meetings."

Then I said, "Dad, what did the prime minister say to you?"

Dad answers, "He said that he needed our yard for the funeral ceremony of our neighbor, the governor, who passed away that day. He also said to me that many people, including ministers and family relatives were on their way here and that there was not enough room to accommodate this entire crowd. He said he would be very thankful if we fulfilled his request for only one hour. Of course, I gave my consent immediately and they brought tables and chairs. Our yard was as big as half a football field and over 150 people flocked in for the funeral ceremony which was conducted by their Imam with the whole crowd praying."

Dad, "I do not remember this event, but I do remember Sar Yaver, the king's first officer who used to come often to our house, knocking at the door first, and sometimes, you would not have patience for him and you asked us to tell him that you were not at home. What did he want from you that he had to come at least once a week?"

"Whenever he came, we would go out for a walk in the streets and talk about different topics. I would say that he felt bored and was just looking for attention besides fulfilling his wish to have a Jewish friend."

Mom raises her voice, saying, "We have good memories of that country. We knew many good people there and had good connections with them. That is all."

"The *Sukkos* holiday came and passed during the fall and we were heading into a difficult winter. Sometimes two to five feet of snow fell in one day and the feast of *Hanukah* always fell in the climax of the winter. What could I tell about this beautiful holiday, which does not require special preparations like *Pesach* or *Sukkos*? *Hanukah* is the festival of lights, which lasts for eight nights in which we lit a *menorah* every night. Instead of candles, we used oil and a wick since many Jewish families prefer this way of lighting as it is similar to the way it was lit in ancient times including in the temple in Jerusalem. This festival commemorates the victory of the Jews over Greece, after they ruled the land of Israel when they were an empire in ancient times. After recapturing the country, the Jews returned to their temple in Jerusalem and renewed its ritual services. The first thing they did in the temple was light its big *menorah* with oil. We at home, always celebrated this holiday with enthusiasm and with special festival meals every day. This winter was terribly cold, but at home we felt warm and comfortable. We had a fireplace where Dad always made sure there was wood. He did the same at the nearby synagogue where he went in the early morning to prepare the heating for the convenience of the people who used to come for the daily morning prayers.

The winters lasted about three months and towards the end of it, around March, we were in the process of preparations for the upcoming *Purim* festival. This is the most amusing holiday of the year and is the last before *Pesach*, which comes out in the beginning of the spring. It commemorates the saving of the Jews of Iran in ancient times from the king's order to execute all of them. That king, named Achashverosh, was a drunkard who approved every evil idea of his chief adviser, Haman. During the time of the issue of this horrible decree, he fell in love with a beautiful girl named Esther. She told him she was Jewish and persuaded him to cancel his decree and to execute Haman. After the king realized what devilish advice his adviser gave him, he ordered to execute, by hanging, Haman and his ten sons. That day was on *Purim*. Since then, we celebrate this holiday with much happiness and joy. On *Purim*, we in Afghanistan used to eat a lot of different types of cookies and sweets, which the women prepared, like *likah*, *sambosa*, *halva* and more. The children used to wear costumes and to hold noisemakers which were used altogether at the mention of Haman's name in the synagogue during the reading of the *Megillah*. This *Megillah* is the story of the rescue of the Jews from Haman by Esther. It is also called the *Megillat Esther* and is read with melody. Each family sends portions of cookies and sweets to their friends and throws parties at their homes. This is a day that even adults have fun and therefore, they allow themselves to play a card game called *chollos*, which involves gambling and money. The children played the same game but instead of money, they used bags of walnuts. We, the kids in Kabul, used also to play in the street with kites. There was a competition among us to see whose kite was more beautiful and could fly the greatest distance. Dad told us that in his youth, there were no such games and he did not like us playing them because of the tough competition.

CHAPTER 8

The Jewish community of Kabul was the biggest in the country. Its members were very organized and took part in all the events of the community, such as celebrations and parties. The invitations were handwritten by the person throwing the party mentioning the date, time, and location. This was enough for all the recipients to come. Urgent events were notified in the synagogue. Everyone, with no exception, took part in them. If someone was missing, people would be sent to check if everything was alright with him or her. Even if someone was missing on a Shabbat, even during prayers, a representative of the synagogue went to his home to check. Parties of *Bar Mitzvah*, circumcision, or weddings, would be done at home, and during summer, in the yard. There were no catering companies and all the meals were prepared at home. The men would buy the food and the women would cook it. Some parties contained as much as 300 people. The first dish was always fish, which were in surplus in Kabul. They came from the river across the town. The main dish was *kabab roqan*, with rice, and for appetizers, we gave out *yachny*.

It is interesting that when war began in the seventies and until today, the Kabul River is almost dried out and fish can hardly be found in it. More severe tragedies and disasters, with no end, were inflicted on this country since the Jews left when the revolution started.

Dad says, "Everyone should believe that Jews bring blessings to the country they live in. There are even Muslims that believe this. I heard it from them personally. How could such disasters happen to a quiet country with pleasant and generous people? Everything is from G-d. We lived together with the Muslims like cousins. You remember it well, I do not have to tell you. You left Kabul when you were seventeen and you remember the favors they did for us."

"Yes, Dad. I remember well. In 1954, I went to a public kindergarten. The first day was very difficult for me. Dad just left me there and told me he would pick me up in the afternoon. I remember that everyday, we got a glass of milk and once a week, a multivitamin capsule to take. My older brother, Amnon, resisted going to the kindergarten. He would cry and stay at home. I attended for a full year. The next year my father sent me to a children's *Yeshiva* managed by *Mollah* Aqajan Basal. At the same time, he registered Amnon and me to a public school to learn general studies. I always went to school, while my brother would sometimes go and other times run away. When Amnon was seven years old, our parents decided to send him to study in Israel with a group of Jewish children. This program was organized and managed by *Aliyat Hanoar,* the youth department of the Jewish Agency. My older sister, Esther, who was twelve years old at the time, decided to join him.

I clearly remember one morning, Mom woke me up quickly to say goodbye to my siblings who were traveling by plane to Israel. The whole family accompanied them to the airport and when the plane took off, Mom cried terribly. Dad calmed her down by saying that in Israel, they will get a great education besides being in good hands, since our grandparents had immigrated to Israel some time before. Although my parents were very active in the community, they always found time to raise us with a lot of love and warmth and to give us a good education.

Amnon's departure caused me a lot of grief since we were very close and it happened so suddenly. I felt lonely. I continued my studies half the time in a public school called Masaod Saad and the other half in the *Yeshiva* for holy studies which was at *Shahar Konah* in the *havli* Haji Yitzchak. It took me a half hour walk to reach each of these schools; one in the morning and the other in the afternoon. The way to the *Yeshiva* in the afternoon was not easy since the children of a police officer who lived there used to bully us. They hit, chased, yelled, and threw stones at us until we reached the yard of the *Yeshiva*. However, when we reached there, we encountered the guard dogs of the *havli*, who chased us until we reached the building. At the *Yeshiva*, we were scared of the *Chalifa*, who used to hit us with his stick. We had to pass these three obstacles every day until we got back home safely.

After a short period, this teacher was replaced by Tzvi Basal, who was only eighteen years old but very smart. He was tough and managed the class successfully. There were always unruly and insolent children, but he controlled them by hitting them with a small stick. He also had a *flack*, a thick piece of wood about a yard and a half in length. He used to tie the feet of a misbehaving boy to the edges of the wood, which prevented the boy from moving, and then he would hit him. We, the kids, did not like this sort of punishment and one day, four of the smartest kids in the class planned to steal his *flack*. I was one of them. The leader of this team was Shlomo Nisani and the rest were Amnon, Elimelech Yazdi, and me. The plan was that each one of us would ask the teacher to go the restrooms in a difference of five minutes. Then, I would enter his office which was adjacent to the classroom and quickly move the *flack* outside. Shlomo would take it from me, and Amnon and Elimelech, who were already standing outside the fence of the building, would get it from him and run away. Shlomo and I joined them afterwards at a specific place and burned this damned *flack*. The day after, when the teacher discovered his *flack* was missing, he became very angry at us since he knew that we left the class on the day before. He came up with a proposal that if we returned the *flack*, he would not punish us at all and forgive us. We said that it was impossible since we burned it. We explained to him that this *flack* caused us nightmares and even parents complained to him in this regard. Fortunately, he accepted our explanation and realized that using this device was a bad idea. We got away from any punishment and he returned to using his stick more often.

I was eight years old when one day, my Dad came to school and told the teacher that this Shabbat was his *Chiyuv,* the date his father passed away. He asked him to teach me the proper melody of the reading of the *Haftorah* at the synagogue. The teacher agreed and said that Michael is a smart boy and he could teach him that within a few days. On the upcoming Shabbat, I was quite ready to read the *Haftara* and I did it in an Afghan melody that the teacher taught me. In the evening of that Shabbat, I was the cantor of the evening prayer for the congregation and my parents were very proud of me. At the *Yeshiva*, we learned all of the prayers, the daily and the holiday ones, as well the popular holy books, the *Torah, Mishna*, and *Zohar*. We also learned Shabbat and holiday songs. The class was of 35 to 40 children, under the management of *Chalifa* Tzvi. He taught us the Hebrew language but the letters were different then the ones in use today. It was the letters of *Rashi*. As part of our exercises in this study, we used to write long letters to our relatives.

During winter, we had a long vacation from the public school so we spent the whole day in the *Yeshiva*. It was really difficult for us since we had only a one-hour break in the afternoon which we spent in the yard of the school. We ran wild in the yard but were forbidden to play in the snow since it was dangerous. We were also forbidden to play *barf jangy* in order that we not get injured. Another restriction inflicted on us was not to play *gody pran* since we used to climb onto the roof in order to fly the kites. Many kids fell down from the roof and broke their arms or legs. Nevertheless, we played this game after school hours outside of the school where the teacher could not see us.

The community got bigger and new children joined the *Yeshiva*. The place had become too small, so the management looked for a bigger new building. We moved to *Shahar Naw* and a new *Kalantar*, *Mollah* Yecheskel, was elected by the community. The community council was made up of *Mollah* Shalom Shamash, *Haji* Yitzchak Basal, *Mollah* Shimshom Kashi, and Shmuel Aaron. Later on, in the sixties, the president was replaced by *Mollah* Meir Ovadia and *Mollah* Levi Haji, who came with his family from Hirat.

Our family also grew with the birth of another son, and Dad threw a big party, with an orchestra, for this occasion. For the first time, Dad decided to be the *Sandak* at the circumcision for the newborn who was named Gavriel. For the sons born before, he would give this honor to relatives of the family or close friends, but this time, Mom had persuaded him to take this honor for himself.

My older siblings, Amnon and Esther, had already traveled to Israel. I really missed them, especially Amnon, with whom I played a lot and went to school with every day. My parents were aware of my feelings and therefore, gave me plenty of attention. They bought me whatever I liked. Dad used to accompany me to the public school sometimes, which was a forty minute walk. He asked the *Sar Mahilim*, Aqai Nadi, who was a very liberal person and liked Jews, to treat me nicely since I was feeling lonely.

In school, I was a quiet, neat boy who dressed well. I was quite good in my studies, but not the best, except in English and geography which I liked and received high grades. The teachers in the public school generally liked me. Once the history teacher entered, looked at me, and said, "Kids, today we will study about Judaism." At the spot, I thought to myself, "What is going on? This is a secular school with an absolute majority of Muslim students and staff, and this teacher wants to teach Judaism?" It sounded weird to me, but as the teacher sat down, he asked me to go to the chalkboard and to tell the class about the lives of our forefathers, Abraham, Isaac, and Jacob. I was a shy boy, but I succeeded in this task. They listened carefully and liked my stories. They especially found interest in the story about Jacob's sons, Joseph and his brothers. I told them about the hatred of the brothers for Joseph because their father loved him the most. This hatred deteriorated to selling Joseph to be a slave in Egypt. Circumstances changed and Joseph had become the first assistant to the king of Egypt and simultaneously, his brothers had the need to go to Egypt to look for food because of a deep drought in Israel. This episode of me becoming a 'lecturer' for Judaism returned several times during the school year.

Once during a class of *Koran*, which Jewish children were exempt from attending, I played in the schoolyard. Suddenly the teacher, named Aimaq, left the class, approached me with a big smile, and shook my hand. That was not usual for a teacher to do with a student. He said to me that he would like to come and visit our house. After discussing with my parents, we invited him for the upcoming holiday. He came to us and with tears in his eyes and a choking voice, he told my parents that his grandmother is Jewish, and her name is Devora. My parents reacted to this revelation with disbelief because his origin was of North Afghanistan. They listened to him but did not say anything, since they did not want to get into trouble. He came to us on several other occasions until he was removed from our school to teach in another city.

Once, in the early sixties, the principal of the public school, Aqai Nadi, entered my class and called me out. "Listen *Yehudi Bacha*, tomorrow morning, dressed beautifully for school. We are going to the palace to welcome the king." I answered excitedly, "Yes sir!" I came home and told my father what the principal wanted from me. He answered, "Why not? What is the problem? Dress in your Shabbat clothes and go. Just be careful not to get dirty." The next day, I woke up in the early morning, prepared myself properly and went to the principal. He took me together with other five children dressed beautifully to the king's palace. We went on foot for forty minutes until we reached the main gate of the palace. Another one hundred children were standing on both sides of the road leading to the palace gate. Suddenly the principal disappeared and after a few minutes, he returned with a package of two types of flags, the Afghan and American flags. Half the children received the Afghan flags and the other half got the American flags. When I got my own flag, it was Afghani, but right away, he took it back and switched it to an American flag saying, "You, the Jews, love America." I was holding the American flag with pride.

I did not know what all this was about when suddenly we saw police-motorcycles and behind them, a beautiful black Cadillac. The convoy came to a stop right at the gate where we were standing. Then we saw the president of the United States, Eisenhower, and the king, Zahir Shah, exit the Cadillac and stand right next to us. They looked at us with wide smiles and entered the gate of the palace. What we saw lasted a few minutes and we returned right away by foot to school. In the evening when I returned home, I told my parents the whole story and they were very excited.

Since then, each year, until the end of sixth grade when I finished Masod Saad elementary school, I was among the chosen ones to represent the school on the Independence Day of Afghanistan. We were taught the *hatan* dance which is a popular Afghan dance with a special costume. On each Independence Day, which always came in August, we the 'chosen children' danced this dance at the opening of the eight day celebration at the national stadium in Kabul. The stadium was totally crowded and there were sport competitions followed by fireworks every night. During these eight days, there were performances of music and entertainment from the best artists in Afghanistan. There were also exhibition booths with exhibits of many foreign countries, with the biggest one from Russia. Most of the imports in Afghanistan came from that country. Even though the United States had tried to put a foothold there, Afghanistan was still a pro-Soviet country and therefore, the help that America could offer was restricted. America sincerely wanted to help this country, as she did for all undeveloped countries, mainly in construction and education.

I remember that during the sixth grade, America built a highway between Kandahar and Kabul, which looked different from all the other roads in the country which were built by Russia. America always sent teachers to teach the English language in the public schools. The American embassy, which was a five-minute walk from my home, also had a cultural unit which was open for the public the whole year. I took different courses there and went to see American movies twice a week. We were always welcomed nicely there. I cannot forget the huge party the American embassy threw for the Afghan public on each New Year. There were shows and movies and at the end, everyone, children and adults, left with valuable presents. America is really a country which offers love and kindness to every country in need. No wonder people say "God Bless America," and indeed, it is a blessed country.

CHAPTER 9

We received letters from my brother, Amnon, and sister, Esther, who were living in Israel with our grandparents. They used to write that everything was okay and that they were getting used to the new country. My duty was to read these letters. Mom always missed them and was crying while Dad did not show any emotions as he internalized them. He was full of hope that in Israel, they would get a better education. At that time, Israel was involved in the first big war after the Independence War. It was the 1956 war with Egypt, who threatened to block the Suez Canal from ships from western countries. Israel fought in the Sinai Peninsula in alliance with Great Britain and France, and won. In Kabul, there were some demonstrations against the state of Israel and fanatic groups harassed the Jewish population. The Afghan king, who liked the Jewish minority sent police to all Jewish neighborhoods in order to secure their safety.

I remember well the neighborhood where we lived. It was in *Shahar Naw* near *Baq Omomi*. It was a beautiful and prestigious neighborhood and almost all of the Jews live there. We were very united and Dad, as always, took care of the community.

Among his duties was matchmaking until it was passed over to the community Rabbi *Mollah* Yosef Aaron. Matchmaking in the community was not an easy task but a rather complicated one. The first one to raise the idea who may fit for another was the Rabbi. If he did not succeed to convince the would-be couple once and twice, then the prominent people of the community would approach them to convince them. If they succeeded, they gave candies to the would-be couple which was a symbol of good luck. Then the process of unifying them began. First the engagement ceremony came, which was exciting. The groom's party brought *Haft Lali,* which are seven big trays filled with sweets decorated with seasonal fruits and dried fruits. Jewelry for the bride was given and a date for the upcoming wedding was determined. Right before the wedding, a short ritual was done for the bride, a *Hana Bandon,* which is hair coloring with a natural reddish brown dye called henna. Then the wedding came and right after that came the ritual of *Sheva Brachot*, which was a sequence of seven blessings' ceremonies occurring on a few consecutive nights at the parents of groom's or the bride's home and included blessings for the couple. On the last night of this ritual, all the guests who attended the wedding were invited to the bride's home and after the ceremony. That night, the groom would take the bride to their new permanent home. Sometimes, if the couple could not afford to buy or rent a new home, then they moved to a section of the groom's parents' home which was reserved for them.

Our family, as usual, got bigger, with a new brother joining us. At that time, Dad was in Business with a partner named Shmuel Aaron, who was one of the richest businessmen in Kabul. He was a type of person who liked to enjoy life and was among the first ones in the country to drive a luxurious car. I remember it was 1958 when Mom gave birth to my brother, Raphael. My older sister, Bracha was sixteen years old and she ran to Dad's work to bring him the good news. I was outdoors, and saw my sister run along with another woman. It was customary for the community that the first one who brings such news gets a present, a garment or suit. My sister was the first one to reach Dad and later got her present. I was still outside when I saw Dad together with his partner arriving home.

Dad started immediately to organize the *Brit Millah* ceremony which took place eight days later together with a big meal for many guests. The event was in the big synagogue and the *Sandak* for the newborn was *Mollah* Shmuel Aaron. As a very rich and generous man, he brought a singer named Challo, who was one of the most famous singers in Afghanistan at that time. This event was one of the greatest that I remember in my life. We were already a family with many children but Mom raised us wonderfully and with devotion. All of the children were taught discipline and respect for the parents. Each one knew his or her duty. Mom was the housewife, and Dad was involved in his own business along with public assistance for the community. When Dad returned home in the evenings, we children were expecting him and sitting near the table. After Dad took a shower and changed, he joined us at the head of the table and instructed us to serve dinner. My sisters placed a *sefre* along with the dishes and bread. We were sitting in order of our ages. Thus, I, as the oldest son in the household, sat next to Dad. We always started this meal with vegetables, usually cucumbers which were very tasty in our country, or lettuce. We used to eat quietly without talking and with no rush. Sometimes, my smaller brothers had no appetite or did not like the food that was served, so they did not eat. No-one forced them to eat if they did not want to. The food on the table was quite enough and more or less according to demand.

Food was never thrown away in our house. If any food was left from the meal, we gave it away the next morning to people who were usually non-Jewish and knocked at our doors. Even crumbs of bread we did not throw but instead, we collected them and put them outside in our yard for the birds. Besides food, which was strictly forbidden to throw away, we were taught not to throw anything on the floor or on the ground in order not make a mess. Also at school, the awareness for keeping the environment clean was high and in case some kids threw papers on the ground there were *Chaparasti* who cleaned the ground immediately. I remember an old man, probably over seventy, with a white beard, named Malang. He used to come once a week. At the entrance of the *havli* of the house, he would very loudly bless the owners of the house. Mom would immediately come out and give him a plate full of home-cooked food. He would eat comfortably and then sing a song before he left on his way. I still remember that song since we always used to sing it:

Choda Vanda Choda Vandi Jahani

(Lord, Lord, of the whole universe)
Choda Vandi Zamin Vosomoni
(Lord, of heaven and earth)
Choda Vanda Tupir Hara Beyamorz
(Lord, remember to bless the old ones)
Jvan Hara Bekam Del Rasany
(And to fulfill all the good wishes of the young)

After finishing the song, he would have blessed everyone and leave with a present for his way. Usually, this was a package of basic food staples like bread or sugar. We considered the blessings of this old beggar a nice way to thank the generous Jews on behalf of all poor people that used to come to them for charity. We also believed that those blessings had a power to help us. I remember another example of someone that always visited the Jewish neighborhood with a monkey on a leash. He came into our yard and performed a show with the monkey in order to get some money. The monkey was friendly and his master trained him to climb on and off the trees by order. When he accomplished his task, he used to smile and laugh at the people, satisfied with himself. His trainer also instructed him '*Shadi bazi mekona*,' which means 'Monkey, dance before them,' and the monkey immediately started to dance while laughing. This show would last about five minutes and we would give him at the end some money as a token of gratitude.

We always used to give to the poor who knocked at our doors either food or money or clothes that we no longer needed. Despite the presence of rich, natural resources in Afghanistan, most of the people were poor. Many laborers were looking for work and therefore, they were ready to receive any payment. Construction workers were defrauded by their contractors by giving them no basic benefits. They had only one break for meals during the day, in which their employer fed them berries with dried bread. It is hard to believe, but these laborers were very happy with what they got since they were at least employed. Even their work was physically intensive compared to work nowadays. At that time, there were no cranes or tractors or any electric machinery. Bricks were carried by a chain of people and passed from one floor to the other by hand. That is how houses and buildings were constructed at that time. It is very interesting that these laborers were healthy people despite these hard conditions. There is no doubt that the good weather, the fresh water they drank, and the natural food they ate protected them from diseases.

Water was flowing from the snowy north through a river into the center of the city. On its way, the river passed many summer resorts north of Kabul, like *Paghman, Golbahar*, and *Gulbaq* where the high society, including many Jews, used to spend their summer. During the cold winter, we traveled many times to the south of the country close to the Pakistani border, a zone called Jlal-Abad. This place had a pleasant weather during winter-time, similar to Florida and many rooms were available for rent.

As mentioned, the country is rich in fruit and vegetables, which have a unique taste and are a main component of the Afghan meal. Usually, bread and cheese are added to meals which are totally natural. There was a renowned saying: *Nan Panir Vacharboza - Bchor Babin Chi Chozmaze* (Melon with cheese and bread - eat and you will feel great). Genuine natural cheeses made from goats and sheep are world renowned for their taste. So were the dried fruits of Afghanistan which is the number one export of the country. The Jews were involved in these exports, especially in pistachios, raisins, and other dried fruits that cannot be found in any other country.

The decades of the 1950's and 60's were a period of prosperity especially in the capital with its fast growth. The businessmen who were mostly Jewish were very involved in this economic growth. They opened offices and stores in famous commercial centers like *Serai Shazadah* and *Serai Qumi*, and later on in a new commercial center called *Hazar Gul* that was built in the western part of the city. In the center of it, there was a trembling bridge called Pol Larzanak which was scary to pass over it, since it was unstable and very narrow relative to its length. Its length was 150 meters while its width was only 1.5 meters. In this zone were many stores, a central post office and near it the Kabul radio building.

Many Jews found interest in the Afghan music with its special musical instruments such as *Armonia, Tabla, Rebab, Setar, Akardyon, Mandolin*, and more. All of these were unique types of accordions and violins. During those years, the famous singers were Ostad Sar Ahang and Rahim Bachsh, and later on, they were joined by Ham Ahang who was renowned for his great, strong voice and soulful melodies. He was invited to sing in almost all great celebrations or parties that took place at that time. In the late 60's, modern music entered the country and pop bands like Zahir Hovida and Ahmad Zayer were doing performances. Luxurious clubs combined with restaurants opened and the most famous one was Spinzar which was located at the top floors of a skyscraper. This club was crowded and very busy with performances every night. More movie theaters opened and modern restaurants usually opened near them.

Luxurious cars started to appear in the capital like Volga from Russia, Mercedes from Germany, and Chevrolet from America. In regards to cars, the change came very fast, since in the 1950's people still used horses and wagons for transportation. In the 1960's, modern buses were imported from America for public service. As the Jewish community advanced to the upper class, each household had a car with a chauffeur. Many motorcycles could be seen on the roads. Dad and his partners bought a Mercedes minibus with eighteen seats which served their business. He also bought an American jeep in which he used to go to work and back. This jeep would be parked next to our house and we the children, as teenagers, used to steal it for very short rides from time to time. One day, Dad caught us with the jeep and angrily shouted at us, "You don't have drivers' licenses and these adventures could end in tragedy besides ruining the jeep." Since then, we drove the jeep only with the permission of the chauffeur who sat next to us for supervision. Those were very good times for us as youngsters.

I remember the new hospital that the Czechs built on the way to the airport called Sad Bistara. It was almost empty of patients since people in general lived a healthy lifestyle. If someone got sick, a physician was called home for treatment. Only people in emergency cases, like car accidents, operations, or the like were hospitalized.

Within the Jewish community, whenever someone was sick, the congregation would pray for his health and come to visit. Once, my little brother, Ishai, when he was around seven years old, got a terrible stomach ache. We did not know what caused it and when household remedies did not help at all, we called an expert American physician. He instructed right away to hospitalize the boy for thorough exams. The whole congregation was alerted. X-rays showed that Ishai had appendicitis and needed immediate surgery. During the operation, Mom, Dad, and a few of their close friends stood near the operating room, praying, while the whole congregation enquired to the status of the boy and wished the family well regards. This was a sample of the behavior of the community in such cases as if something extraordinary and complicated had occurred.

My brother Gavriel, at the age of two, became sick several times because of human errors. Medication meant for his ear was given for nose mistakenly and caused severe damage to his nose. Then one time he suffered from an infection of throat, nose and ear. A physician came to our home and prescribed him medications. My older sister accidentally switched between them and the results were traumatic. My brother cried and screamed all night until the early morning. He needed harsh treatment, which left him with difficulties in speaking until he grew up.

I was still attending two schools daily. In the morning, I went to public school to learn general studies and in the afternoon to the *Yeshiva* to learn holy studies. I did not miss a single day and I always came on time. Winter time, when I studied only in the *Yeshiva*, was most difficult for me since the road to school was covered with deep snow, almost a meter high. I had to go by foot and it took me approximately an hour to reach school at *Shahar Kohna*. Once, the weather was very stormy, and to my surprise, only my teacher and I came to the *Yeshiva*. He shouted at me how I could have come on such a day and he asked me to come close to the heater to warm up and then to pray the midday prayer and go back home immediately. After spending an hour there, I went back home, again by foot, for another hour. I enjoyed letting the beautiful snow fall on me in great quantities and seeing everything covered in white, the trees, roads, fields, roofs, and more.

In those days of heavy snow, many laborers looked for cleaning work at private homes. At our house with a roof covering a dozen rooms besides a huge yard, there was plenty of work to be done to *barf pak*. It had to been done right away, otherwise the soft snow turned to ice, and there was a danger of slipping. Besides, if the sun appeared and the weather got warmer, the ice would melt and drip over the roof and even into the house. We the children, together with Dad, cleaned the yard joyously. This deep snow was really an unusual adventure which I cannot forget.

On free days from school, like Saturdays and holidays, four of my best friends and I gathered and played together. These friends were Weizman Shamash, the son of *Mollah* Shalom Shamash who was the leader of the community, Binyamin Moradoff, Amnon and Elimelech Yazdi. We were very close friends and on heavy-snow days, used to build huge snowmen. All of us were more or less the same age and started, at the same time, to put on *Tfillin.* This item was very expensive, worth a few hundred dollars and Dad ordered it for me a long time in advance from Israel through Grandpa Yehuda.

In the year 1960, my brother Amnon reached the age of thirteen, the time of *Bar Mitzvah.* He along with our sister was already in the Holy Land for sever years and kept us updated with frequent letters, usually containing good news. Dad sent a lot of money to Israel to make Amnon a nice *Bar Mitzvah* party. We often received letters from Grandpa and Grandma urging us to leave Afghanistan and come to our Land of Israel. They asked us over and over how much longer we would stay in the Diaspora. They wrote, "Please come to the Land of Israel. It is a beautiful flourishing country." Dad answered, "It is not yet the time for us to come. The Jewish community here still needs my help and when the time comes, we will join you."

As mentioned, at that period of time, businesses in Afghanistan were at their peak and Jews were very involved in this economical success. Many of them made plenty of money but at the same time, they kept to the religion. The Shabbat and holidays as well as kosher food were strictly observed and above all, they were aware of being united as a minority in a foreign country. I would say that this unity was the main cause of their success at business. The Jews were very generous and friendly to others which helped too.

In our house, I have never heard curses or obscene words. Sometimes we were slapped or pinched and the worst was to say to us, '*Misel Chodet Psar Kony*, May your child would be like you,' if we misbehaved. If we were good, and helped our parents or our relatives voluntarily or upon request, we were right away blessed by them. Such blessings were like '*Peer She, may* you make it to old age', '*Safid Bacht She*, may you have a lot of luck', and others like this. Even when we left home to school or to the playground, Mom would bless us, '*Choda Hamrat, Malach-neck Hamrat* May G-d and his angels protect you.'

CHAPTER 10

During those years, Dad traveled away from the city for business often, mostly to the city of Qonduz, which was a ten hour drive from Kabul. Qonduz was a city known for its fashion stores. Their merchants used to come to Kabul to buy the raw material on credit. Dad sold them the merchandise and afterwards needed to go to Qonduz to collect the money. Once while Mom was pregnant in her ninth month, Dad left on business and said that he would come back right away. Somehow, he was delayed and the children (I was 12 years old back then) had become worried. However, Mom was not nervous and was absolutely sure that he would be back very soon. She said, "Apparently, he did not accomplish his business yet and besides, it is not the first time such a delay happened." And she continued, "I gave birth 10 times already and I do not need any help. Everything will be okay." While we were talking, Dad had arrived. We, the children, were so happy. Dad, as always, brought presents for us with him. This time, he gave each of us candies and *moza* for the upcoming winter. We were very happy with them since they were covered with fur inside and the outside was made from beautiful, shiny rubber. We also knew that Dad came back with a lot of money that he collected from the merchants and we would feel some change in our lifestyle.

The next day, Mom said to Dad, "You really returned just on time. You should go call the midwife now." This happened during an evening in the month of November when Dad went out and returned with the midwife. She went right away into Mom's small room. Also inside was the wife of Rabbi Yosef Aharon, who was a very close friend of hers. Dad and the children were sitting in the living room next to this room. We heard a few small screams from Mom and soon after, the crying of a little baby. The children were very nervous, and so was Dad, who was leaning his head on the wall. We didn't know if it was a boy or a girl. Suddenly, the door opened and the Rabbanit came out yelling loudly "*Mazal tov*! It is a baby boy." What a joy it was especially for the boys who wanted another brother to join the family. Our sisters would have preferred a girl…

Dad raised his eyes to heaven saying, "Thank G-d this has passed. This is my youngest and tenth child and I am really happy and proud." Dad wanted to make the *Brit Millah* party within the synagogue which was adjacent to our home. On Shabbat, many guests came to honor Dad as is customary. A name had to be chosen and Dad came up with several suggestions which Mom did not like. However before that, a *Sandak* for the newborn had to be chosen. Mom and Dad together agreed upon to give this honor to one of the most prominent members of the community, *Mollah* Haji Yitzchak Basal. This man was very smart, and a wealthy businessman who gave a lot to charity. He was very happy to be chosen and Mom consulted with him regarding what name to give the baby. He responded that on this Shabbat, the Torah chapter to be read would be about the patriarch Avraham. Thus, he thought it would be very appropriate to name the boy Avraham. Mom and Dad liked this idea and accepted his suggestion.

Mom raised Avraham as enthusiastically as if he was her firstborn. Most of the time, the baby slept in the *gvare*. There were no diapers in Afghanistan then, but we had a special device installed beneath the bed to clean the waste. I had already been experienced in raising little kids and I held him often in my arms and played with him. When I was at home, I always waited for him to wake up so I could play with him. I loved him very much, and I still have special feelings for little children today. In his first year, Mom nursed him as she did with her previous children. Avraham grew up to be a beautiful, charming boy.

The community grew very fast, especially with new children. There was a demand for a new *Yeshiva* which was established in the synagogue. This synagogue shared a huge yard with our house. Since that *Yeshiva*, under the leadership of *Chalifa* Tzvi opened, the yard was always crowded with people and children. Actually, the huge building of the synagogue served also as a community center and every event regarding our religion took place there. For example, gatherings of the community, preparing boys for *Bar Mitzvahs,* preparing couples for their weddings, as well as preparing kosher food, like slaughtering chickens and wine production according to Jewish Law.

The teacher Tzvi, who was a very active man, got a separate room for himself at the entrance of the synagogue complex which served as his office and study room. He helped Dad tremendously in the daily preparations of the synagogue for the morning prayer. For this, it was necessary for them to wake up early in the morning. After the prayer, children came to the *Yeshiva* for their holy studies until four in the afternoon. The children always studied out loud since this was a better way to remember the material and to learn faster. Mom, who usually was at home during these hours, loved to hear the voices of the children studying Torah.

I was among the older boys of the *Yeshiva* attending the last grade. I was almost thirteen, the age of *Bar Mitzvah* and had to be prepared for it. I had already gotten my *Tfillin* from Grandpa Yehuda in Israel. He enclosed a letter with them with blessings for success. The teacher privately taught me how to put on the *Tfillin* and what blessings had to be recited. In August of 1961 my parents scheduled the date and time for my *Bar Mitzvah* party. No invitations were sent out, but Dad announced in the synagogue the event to come and invited the whole congregation with no exception. He did the same at work. I invited all of my friends as well as my roommates from the *Yeshiva*. Dad and *Chalifa* Tzvi, who had become a close friend of his, prepared the huge yard for the *Bar Mitzvah* party. Benches and tables were spread over the lawn and in the center was the platform which was placed with a table and chairs for the honorees.

The table was covered with a tablecloth and beautiful flowers in vases. Dad installed wires with light bulbs throughout the yard since the celebration was scheduled for the evening. He even invited a professional photographer which was very rare to do for such a party. Many guests of the city of Hirat were invited too.

The happy event was on a Tuesday evening before dark. It was customary for the father to buy the first watch for the *Bar Mitzvah* boy which I put on very proudly that evening. Each of the invited guests appeared with no exception and took their seats. My parents and siblings sat beside me and the teacher, Tzvi, called me to give the introductory blessings. I started, and then heard the teacher yell, "No one can hear you!" There were no microphones there yet, so I needed to read the blessings loudly and the audience responded with "Amen." The ceremony took about fifteen minutes and at the end of it I saw packages of candies being thrown above my head from all directions and people were shouting *Mazal Tov*. I got plenty of kisses from my proud parents. Right after the blessings, the special meal started and Dad and Mom moved from table to table with a bottle of arak wishing each guest personally a warm welcome. The whole celebration ended successfully with much happiness.

The next morning, two of my good friends, Binyamin and Weizman, visited me at home and each of them gave me a white shirt for Shabbat as a present. They wished me "*Mazal Tov*" and I thanked them and wished that their upcoming *Bar Mitzvah* will be as successfully as mine was. We went out for a trip in the city and I noticed that they were staring at my watch in amazement. They said, "We wish to also get such a beautiful watch for our upcoming *Bar Mitzvah*." At that time, most people in Afghanistan did not have watches since they were too expensive for them. They knew the time by signs such as at noontime, a cannon ball would be shot from the top of a mountain which was heard all over the capital. At sunset, there was an alarm for one minute.

On the coming Shabbat, I was called to read the Torah in the synagogue for the first time as an adult. Again, packages of candies were thrown at me when I finished the reading successfully. Since then, I could join adults in prayer because reaching the age of thirteen means having to comply with all Jewish religious obligations. I started to feel and to behave like a real man in all aspects.

I finished elementary public school and it was time to register for junior high school. I had trouble deciding between two schools that seemed appropriate for me. There was one that had just moved to a new building called Lesai Habibya. It had a reputation for having good teachers and management. At the same time, another junior high school opened called Lesai Nadirya. My principal from the elementary school, Aqai Nadi, recommended to my Dad to register me to the second school. He told him that its principal, Aqai Qrar, is an honest person with excellent credentials in education besides being sympathetic to Jews. I went to that school with a referral from my previous principal. The registration was over already but nevertheless, I was admitted to the best class out of three for seventh graders.

I was really happy since I was the first Jewish student in this school. Jewish boys attended other schools like Istiqlal, Habibya, Nejat, and Ghazi. All of them were schools for boys since gender separation in the schools was the rule in Afghanistan. Actually, all public places in this country were separated between men and women such as swimming pools, rivers, and even parks. The women were not deprived of equal rights and had there own places to attend. My sisters went to a school for girls called Lesai Malalai and used to hang out in a special park for women called Baq-Omomi Zananah. During summers, I used to go with my schoolmates to the river, Band Qarqa, where I learned to swim.

My parents observed my development and noticed that I had matured more. Mom called me one day and said, "Listen Michael. You are a big boy already. Why don't you pick one of the vacant rooms in our big house for yourself? It does not suit you anymore to sleep together with your older sister in one room." I instinctively responded, "Mom, you are absolutely right. It is time for me to have privacy."

The next day, I cleaned and polished one of the vacant rooms and placed a nice carpet, which Dad gave me once, on the floor. I moved in a bed, table, and chairs and placed some nice pictures on the walls and the room became beautiful. I even called an electrician to install lights above my head. This had become my first private room. Until then, I slept in one room with my older sister Bracha, for protection. At the time, scorpions had infested the city and houses were in danger. One night towards dawn, my sister woke up crying from terrible pain. A scorpion was crawling on her bed and had stung her. I jumped off the bed and called Dad who came in immediately and killed the scorpion.

I spent more time in my private room since I really liked it. Apparently, my family members liked it too. Dad used the room to welcome important guests. My friends came to this room to study together and prepare homework. Even my chemistry teacher sometimes came to me to check the exams of his students since he felt great privacy here.

One day after a period of time, I saw another bed in my room. My sister, Tzippora, had decided to settle there. I resisted, but she was very stubborn and insisted very strongly. I had no choice but to give in. One night while both of us were lying in our beds, we suddenly saw a huge scorpion on the ceiling. We were very frightened. I instinctively took a shoe and got on Tzippora's bed in order to kill the scorpion. Before hitting it, I became frozen with fear that I would miss and the scorpion would fall on me. The same thing happened to my sister who tried to help. I ran to wake up my cousin, Chai Cohen, who had come from Hirat to visit us. He was surprised and asked what was going on. I told him about the scorpion and he told me to kill it myself. I explained to him that I was too afraid and he laughed loudly. He came into my room and the scorpion was still at the same spot. One bang with his shoe and the scorpion was dead. We were all able to breath easier but my sister and I could not sleep that night.

In school, there were a lot of scorpions crawling about. Children in the class were amused by the scorpions and liked to play with them after they cut off their stingers. The scorpions would crawl up and down their arms very quickly. Other types of unforgettable adventures, that were no less scary, were the frequent earthquakes.

The country is located in a geographic area with a high risk for severe earthquakes. When they happened, we would jump through the windows into the yard or find shelter beneath the door frames. I will not forget an earthquake which happened in the middle of a wedding party. The whole building trembled and most of the guests jumped out the windows. Fortunately, even those who were frozen in fear survived with no injuries.

Wedding of my close friends Ben-tzion Kashi and Shoshana (Shamash). The first wedding I remember was a community celebration, and as is traditional in Judaism, it lasted seven days (winter 1955).

Members of the Jewish community of Kabul during a summer retreat at Paghman Resort House, north of Kabul (summer 1953).

Molla Haji Itzhak Basal (right), one of the wealthiest Jews in Kabul with his Workers and truck load of Afghan carpets awaiting export. To his left is his partner Molla Haim Gol. (1958)

Left to right: Aq jamshid Kashi, Mulla Yechezkel Kalanter, Mulla Shalom, Shamash Abraham, Rabbai Yosef Aharon

Left to right: Mulla Elyaho Ambalo, Mulla Shimon Gol, Mulla Levi Haji Nissani, Mulla Shamuel Aharon

Left to right: Mulla Babajan Gul Vardi, Mulla Nissan Yazdy Shimoni, Chlifa Tzvi Bassal, Mulla Avrashk Aharonoff

Community Leaders on Kabul 1950's–1960's. These individuals served as the Jewish Liaisons to the Government and helped fuel the import-export business of the country.

Myself in 1959. I was chosen along with other youth members to represent Afghanistan during its Independence Day celebrations. I was also chosen to make a presentation to King Zahir-Shah and President Eisenhower in an historic visit to the King's royal palace.

My Bar-Mitzvah Celebration (Summer 1961). Picture of myself with my family and my teacher Chalifah Tzvi reading scriptures for the Bar Mitzvah Ceremony.

CHAPTER 11

The year was 1962. I was considered a man already and people in the community approached me with demands, especially since the synagogue and *Yeshiva* were adjacent to our home. Dad or the teacher, *Chalifa* Tzvi, would wake me in the early morning to join the group for the morning prayer. I was required to recite the Psalms in front of them daily. The teacher Tzvi asked me to stay half of the day in the *Yeshiva* in order to assist him in teaching. I did this after coming home from public school. During vacations, I spent the whole day in the *Yeshiva* studying and teaching. I taught the younger children about the Jewish holidays and their customs. I also translated for them religious stories from Hebrew to the Afghan language such as the *Megillat Esther* or the *Haggadah.*

In the synagogue, some members saw me as mature and capable to assist in their offices. They approached my father with a request to stop my studies and be available to help them. Dad responded that it was too early and I had to study. He then asked for my opinion and I answered him that this year I was going to graduate from the *Yeshiva* so that I can be available for half a day for work.

I had been admitted for work at the office of a rich businessman, *Mollah* Eliyaju Amballu. I worked there for half a day every day, preparing his office in the morning before he came in. He handled imports and exports, mainly with the U.S., Japan, and India in the fields of automobile parts and linen. Later on, he went into the business of real estate. I loved to work with him since he was smart and humble and I could learn from him how to manage life properly.

Dad left the business partnership with his friends and opened an independent business of his own. He asked me to work with him for half the day and the other half to study. When Dad asked anything of me I had to listen. I was already in the eighth grade. I was very enthusiastic to work and curious to know how Dad's new business operated. I knew that the turnover was less than Amballu's, but most of the sales have been done on a credit basis, which meant I could assist in collecting the money, and I liked to travel. Exactly at that time, my eldest sister, Esther, who lived in Israel since 1953, got a match offer with an Afghan man of the Simantov family. They sent messages and letters to Dad to give his consent for the match as it was customary in those days. Dad responded that he was giving the full rights to decision-making to Grandpa and Grandma for this purpose, since Esther lived with them in Tel Aviv. He would accept what they would say. The decision was positive and my sister got married to Levi Simantov.

A few months later, my other sister, Bracha, got an interesting match offer. The teacher of our *Yeshiva*, Tzvi, who was like a part of our family, wanted to marry her. He sent the head of the community, *Mollah* Shalom Shamash to my parents with the proposal. Dad and Mom objected at first saying that soon we would all be moving to the land of Israel and the marriage would have been inconvenient at that time. Tzvi was a persistent man who did not accept 'no' for an answer. He sent more and more messengers to convince my parents to change their decision until he succeeded. Dad had been convinced since Bracha was old enough to be married; and if Tzvi, whom we knew very well as a nice and smart man wanted her so much, then it was a sign from heaven that they would be a good match. What was left for us to do was to ask Bracha for her consent to marry him, which she gave. Then we needed the consent of Tzvi's parents, who lived in Hirat, and a week was given for this as was customary in the community. This would occur in a special gathering of the closest family members together with the most prominent members of the community at the home of the future bride. His parents notified us that they were giving their consent but they could not afford to come to the gathering and therefore, *Mollah* Shalom Shamash, the head of the community, would represent the future groom.

At the scheduled date, the ceremony started with the attendance of the community Rabbi, *Mollah* Yosef Aharon, and the rest of the guests. An official declaration of "*Mazal Tov*" with *shiriny* for the groom was supposed to finalize the consents procession. Dad sent me to the adjacent room where the women were attending to take the *shiriny* from Mom. She gave me a huge plate all covered with a variety of candies and I brought it into the men's room. Everyone took from the plate while the Rabbi took a bunch of candies and placed them in his pocket to pass over to the future groom. This was the custom of the Jews in Afghanistan from generations of the past.

A week later, a huge party to celebrate the engagement took place in our house. The side of the groom-to-be offered to the guests seven huge trays covered with many types of fruits, dried fruits and whole chickens. Presents, including jewelry, were exchanged between the families of the two sides and a date for the wedding was scheduled.

The religious ritual of the wedding was short. Four people held the edges of a huge Tallit over the heads of the marrying couple and the presiding Rabbi. Boys and girls holding lit candles were standing around while the Rabbi loudly read the *Ketuba* and blessed the couple over a cup of wine. This ceremony took about ten minutes and at the end, all the guests shouted "*Mazal tov*." From then on, they were husband and wife. Right after, the newly married couple went into a separate room to break their fast of the day, as is customary. The guests took their seats for the celebration meal and started with the smorgasbord until the couple came out of their room to join them. Then the main meal was served, followed by singing and dancing until the late hours of the night.

My parents gave me the feeling that I was an adult even though I was quite young. I looked and felt like I was 20 years old, but at the time of the wedding, I was only fourteen. My mother approached me with a small glass of arak saying, "You are old enough to drink this." Dad, who was sitting near me, heard this and I blushed. I took the cup from her but put it aside while Dad said,"No, no. You may drink it. It's okay and I wish you will marry a nice woman soon." Then I answered, "Me? What? A wedding?" My parents smiled and said *L'chaim* and I drank an alcoholic beverage for the first time in my life. It was very spicy for me. My Mom noticed and gave me a peeled cucumber. Since then, they gave me every arak every Shabbat.

My father asked me to assist him in his business in collecting money from customers out of the city. Dad was tired from his long-distance travels and he needed someone to replace him. My first mission was to go to the city of Qundos and its surroundings. I prepared myself with a small suitcase for this eight hour ride, which was relatively fast since a new highway with a long-distance tunnel through the Salang Mountains was just built. Dad gave me a list of about fifteen merchants whom I needed to visit. I left on an early morning with a chauffeur of a new Volga car with four others sitting behind me. This chauffeur, Aq Mohammad, was a good friend of my Father's, who used to come to our home after each long ride he drove. He also liked to drink. I was sitting near him feeling self-confident and a huge desire to succeed in my first task. We drove on newly built roads that the Russians had just finished and passed little villages on both sides of the highway before reaching the city of Qundos in the late afternoon.

The chauffeur accompanied me into the motel and helped me to rent a room. He instructed me to take a shower and change my clothes. He said to me, "I will come back in an hour to take you to the governor of the city to say hello." I organized the contents of my suitcase, took a shower, and ate the *yachny* Mom gave me for the trip. After I had finished a bit of my meal, the chauffeur arrived and I had to leave right away with him. After a short drive we reached a developed and beautiful neighborhood and he stopped the car at the gate of a new, giant villa.

The guard let us in and I saw the governor standing at the front door of the house waiting for us. As we got out of the car, the governor welcomed us with a smile and the chauffeur introduced me to him. He said, "Please come in and make yourselves at home." We were served tea and candies and the governor enquired as to how my family was doing. Then he said, "You may stay here during your visit to the city. It is no trouble since we have many vacant rooms in the house." The chauffeur responded immediately, "No your honor, I already took him to the motel according to his father's instructions." The governor shouted, "Why does he need a motel when such a house is offered to him?" Then the chauffeur told him about the discussion that took place at our home before we left. "His father did not know whether to send him to you or to the mayor," the chauffeur said. "So he instructed me at the end to take his son to a motel, thus not causing any inconvenience for you."

The governor accepted this explanation but asked me to come over to him before leaving the city. He added, "I want you for a whole day to show you the city and especially to tour a new factory that we just finished building which manufactures oil from cottonseeds." I answered, "That's a great idea. My father told me about this plant many times. I am especially interested to observe the manufacturing process of this type of oil, since cottonseeds are considered Kosher, and the Jewish community could use it for cooking. As of now, we only use home-made oil, which the women laboriously prepare from sheep fat."

The whole week, I was busy collecting money which merchants owed my Father. I started in the city of Qundos and then moved on to the nearby city of Chan-Abad. In the evenings, I returned to my room in the motel. I had collected a large sum of money and guarded it the whole time, which prevented me from leaving the motel at night. On Friday and Saturday, I spent the entire time at the motel.

On Saturday evening, at the end of the Shabbat, I got an unexpected visit by the governor's son. He was my age and he asked me to accompany him to his house. I accepted his offer and went there for a few hours. The house had many rooms that had little furniture, but were covered with expensive carpets that were handmade in Afghanistan. The governor's son asked me to go smoke with him. I did not object since I considered myself mature enough, but of course, I did not want my father to know. Children in my class used to smoke cigarettes hiding behind the school building. It was unacceptable for us to smoke in public, in front of the teachers or in front of our elders.

I returned to my room in the motel to get a bit of sleep in the early morning and again went out for my last debt collections. As I promised the governor, I put aside a whole day to make a tour of the oil factory. I called him on that day and he said to me, "*Bro Bachim* to the plant. The manager there is expecting you and he will be your tour guide." The chauffeur drove me over there and we entered the huge complex through the back. A man about thirty years old approached me with a friendly welcome and introduced himself as Davod. I said, "Nice to meet you. My name is Michael." He spoke to me with a lot of respect, and I felt uncomfortable since I was just a teenager. After a short introduction, he immediately started the tour.

First, he showed me the machine in which the raw material is poured in. Then we moved to see other machines that continue the process of transforming the cotton into seeds by removing wastes. Then another machine extracted oil from the seeds into a huge barrel. The last machine stored the oil into cans. The whole guided tour took about two hours. We came out at the front part of the building and he gave me a big can of oil as a present for my parents in Kabul. I was a bit exhausted from this long tour but felt very satisfied.

The chauffeur took me back to the motel where a few local high school students were waiting for me. The governor's son had sent them. They said, "We have come here to take you to a musical show organized by our school where you will meet new friends." Since the concert was scheduled to begin very soon, I had only half an hour left to eat a bit and to get dressed. We drove to the school where at the front, a huge, decorated platform was installed. Seated there were the parents of the students waiting for the show to start. I met there the brother of my school's principal, Abdul Nadi, whom I knew from before. Finally, the concert started and lasted until dark. I enjoyed it very much and I returned to my room tired.

The next morning, I was scheduled to return home ten days of my first long-distance business trip. I had never been so far from home before or for such a long time. People treated me very generously as if I was a family member. They never gave me the feeling that I was a foreigner or someone not of their religion. I would summarize their attitude to me in two words: respectful and supportive. My devoted chauffeur Aq Mohammad and I left town en-route to Kabul on the same beautiful paths and exciting landscapes that had taken us to Qundos.

I waited eagerly to arrive home to tell my family all the many adventures I had experienced; but above all, I was impressed by the reputation my father had acquired in the city of Qundos and its surroundings. He was positively well-known by all the public figures and businessmen of the city. I really felt proud to discover this. I knew that Dad would not like to hear these compliments from me, but I could not hold back. In the early evening, we arrived home and the chauffeur helped me carry the suitcase into my room. Dad was still at his business so the chauffeur decided to wait for him to come back, which I assumed was for getting paid. Meanwhile, as usual, he spent the time at our home by have one drink after another.

My mother and siblings who had not seen me for ten days, had missed me a lot. I had missed them too. All of them came into my room and bombarded me with questions, 'How was it? What did I do? What did I see?' And more questions. Dad returned home, took a shower, changed his clothes, and then came into my room. "Hi son. *Manda Nebashi*," he said. Then he turned to the chauffeur and said, "Welcome, both of you." The chauffeur, always with a smile when he talks to Dad, said, "Your son is a very nice boy and in my view, he is now considered no less prominent than you by all the people we met. You really have good fortune to have such a son, sir." Dad sat down, had a drink with the chauffeur, and thanked him heartily for the good service and protection that he provided me. Dad wanted to pay him but to my surprise, the chauffeur angrily refused. Dad forcefully put money into the chauffeur's pocket while he was getting ready to leave.

I opened the suitcase and took out all the money I collected with a detailed list of how much each merchant paid. I also took out all the presents I obtained from the trip, some of them were for Dad. I told him that I received a lot of respect and I was aware that it was because of him. "Yes," Dad responded. "Relationships are based on mutuality. Give respect and you will get respect. The people that you met even consider you as part of their family if they respect you." Dad added, "Actually, most of them are Jews as I told you once. They belong to the Jewish tribes that disappeared over time.

Then Dad recalled an event that happened just a few years prior. "There was a big merchant in the south of the country called Chatak. He belonged to the Jaji tribe. He used to send his son to Kabul to buy or order automobile parts and other things from Jewish businessmen. The son behaved very quietly and sometimes, did not talk at all. He would just make the purchase and leave. His appearance was absolutely Jewish in my opinion.

One day, I caught his arm and told him not to be angry at me, because I wanted to ask him something very personal. I asked him if he was Jewish and then he quickly released his arm, told me never to ask him again, and left. After a period of time, he arrived again at the capital for the purpose of making purchases. He came to me and said that his dad wanted to see me to have a very important discussion. I decided not to go by myself and asked three other prominent members of the community to join me on this trip. The three were *Mollah* Shalom Shamash, Shmuel Aharon, and Eliyaju Amballu. We had all done big business with him, but had never seen him. Everything had been done through his son.

We arrived at his home, which was actually a castle and were greeted warmly by him. He was a very tall man, taller than all of us, with bright skin and a long beard. He invited us in and we sat on cushions on the floor. We were served tea and candies and talked only about business matters. After an hour of talking, we stood up to leave. While we were leaving, Mr. Chatak said to his son, 'Who bothered you with the Jewish issue?' His son pointed at me. He instructed me to stay while my friends could leave.

I was stricken with fear and my whole body trembled. 'What did he want from me?' I thought to myself. Only the three of us remained in the room and it was totally quiet. Then Mr. Chatak said,

'Listen my son. This Jew is one hundred percent correct. We belong to one of the lost Israeli tribes and that is why we comply with many of the Jewish commandments. As you know, your mother would light candles for Shabbat and we do not eat meat and milk together. We observe the family purification laws exactly as the Jews do and every newborn boy is circumcised. All of these laws we have observed for many generations as it was passed over from father to son. The only major thing that we differ from the Jews of today is the language which has been forgotten over time.

"I left them with a Shalom." Dad adds, "I had a good feeling that I was right with my guess that they originated from the Jews. I joined my friends who were waiting for me outside and we returned home together. On the way, I told them what happened after they left me alone and they responded that they were also aware of the existence of those tribes, mainly in the southern part of the country."

Dad then turned to me and said, "You are now at the beginning of your mature life and I am sure that you will encounter many surprises."

I answered, "I am sure that you are right since at my first long trip I had already encountered something unbelievable. The generosity and love that people showed me in a Muslim country was an absolute surprise for me. I do not mind to go there again on business missions to help you." My Dad burst in laughter. He liked this idea. I returned to school and was enthusiastic to see my classmates again. One of them, named Yusuf, was my closest friend there. He was like a brother to me. His father was a high-ranking officer in the ministry of education who wrote textbooks for elementary schools. Yusuf visited my home almost every day and we did our homework together as well as take walks in the neighborhood. He was a smart and pleasant guy who looked like one of us.

I was already in the eighth grade and taking part in the soccer competitions among the local schools. My school's team was the leading one with the best players. This achievement was not accidental but the result of nonstop efforts done by the principal, Aqai Grar, who did all he could to put his school in the first place in every contest. Our school always achieved the top place in sports competitions and this was our main source of pride.

I was already considered a business owner who actually ran the business of my father. Still, the turnover of it was low since Dad did not have enough money to buy a bigger business and grow with it. However, he succeeded to provide the family with a comfortable lifestyle. He spent money from his own pocket on community needs. Even though they were individually small expenses, they accumulated little by little into big sums.

I myself received a salary from which I bought a new bicycle. This was my transportation vehicle. At that time, the roads were overcrowded with motor-vehicles and buses, especially during special events such as important games like *bozkashi*. The whole city turned into one huge traffic jam. Still, everything was under control and in order, and traffic policemen stood at every junction directing the drivers. Even the radio broadcasters contributed their share to the safety of the crowded traffic. For instance, every quarter hour the radio news channelon Kabul would broadcast a slogan in the Afghan language:

" Dear brothers, *Az-Narasidan, Der-Rasidan Behtar-ast*, better to arrive late than never!"

Months passed and I continued to work half a day at Dad's store and the other half studying in junior high until I graduated successfully.

Friends of my father advised my Dad to stop my studies since I was educated enough in their view and he might be able to use me full time in his business. Dad did not listen to them and said, "Continue to study and I will get along by myself. I know how to do it." As I knew Dad, he did not have the desire to be a very rich man since he simply did not need it. For him, if he had money to provide a family with food and clothing it was quite enough and any more money was a bonus. He also thought that in any case the earnings of everyone are determined only by G-d and if you put more efforts it will not change the final result. That is how most of the Afghans felt. They did not covet riches.

I got a second chance to go for a business mission to Qundos. The adventures and enjoyments I had were similar to the last time. Dad wanted to send the Kosher supervisor of the community, *Mollah* David Moradoff to the oil plant in Qundos, which I visited for the second time. The purpose of this emissary should have been to make the machines in which the oil is produced Kosher. This means that they had to be cleaned thoroughly in high temperatures under an authorized supervisor so that the oil produced from the cottonseeds will get the kosher seal.

Then Dad realized that in two months he would need to leave for Qundos for business, so he decided that the first visit of the kosher supervisor to the plant would take place on that occasion. Just before the scheduled trip, a few members of the community asked to join in order to see this plant in action. All the machines were cleaned to be Kosher and the supervisor affirmed that from then on, the produced oil was not only kosher, but even Kosher for *Pesach*. The delegation returned home with many cans of oil to be circulated in the Jewish community. A special store was opened for this purpose in the neighborhood of the synagogue and each family was entitled to buy only a few cans until the production at the plant grew. Since then, the Jewish community only used this oil for cooking purposes which was *pareve*.

My friends wanted to hang out with me but I had almost no free time. When I was out of school I worked for Dad. They complained to me, "Why can't you stop working and hang out or play with us?" I told them that I felt sorry, but my Father needed help, so I had to work. Some friends came over to the store and saw me in action. I made sales, counted money, and prepared packages of bills to place in a bag. Most of this money was needed to either pay debts for other businessmen or to be deposited in the bank. The rest was placed in a wooden chest with a plain lock on it. There were no thefts, and no steel safes were used or even available. All sales and transactions were paid by cash since checks were not available.

My friends saw me at work and they seemed to be a bit jealous. "What a good life you have." They would say. "You are running the business of your father. You are a business owner already." "Yes," I answered. "That is why I do not have time to spend with you and I really feel sorry, but I do not have a choice."

There was a youth of the royal family who was a first cousin of the king. We became friends and he used to visit me from time to time. The amazing thing was that he used to travel in a luxurious black Cadillac driven by a chauffeur. He would like to sit in my room and drink and then we would get out for a ride in his car. On Saturday evenings, we used to drive to the movie theater. Older friends of mine had relationships with the sons of the king. The Jewish community in general had close and good relationships with the royal family. On Islamic holidays, especially on Aid Qurban or on their New Year's Day, the leaders of the community headed by the Rabbi, *Mollah* Yosef Aharon went to the king's palace with gifts to wish him a happy holiday.

In the same year, a special incident occurred which could be considered a miracle or divine intervention. The leaders of the community were sitting with the king at the palace and gave him presents. The king was in a good mood and very happy to welcome them. The Rabbi asked how he was doing and the king answered, "*Mollah Sahib*, I have suffered for a long time from pain in my legs, especially during the evening when I am resting. I received all the available treatments and nothing helped. I am asking you, your honor, to pray for my health."

The Rabbi immediately stood up from his seat, blessed the king, and prayed to G-d to cure him. He also told the king that on each Shabbat, during the Torah reading, the congregation prayed for the welfare of the king and the country which is customary for all Jewish congregations where they reside. "With G-d's help you will recover speedily." The Rabbi then left.

The day after, the king took a bath in the palace, placed on his gown, and then sat on his armchair. Suddenly, a scorpion appeared and stung his aching leg. The king rose screaming painfully. His closest aides tried to find the scorpion, and simultaneously called for Karim Mar Gir, who was an expert at trapping snakes and scorpions. However, they found no scorpion. Surprisingly, the king became completely healthy and no longer had pain in his legs. A while later, I met the king's brother-in-law whose name was Sardar Baqi. He verified this story to me adding, "We salute the Rabbi and fully appreciate the Jews' love for the Torah." While saying this, he straightened his body and raised his arm in salute. I will never this episode which brought much respect to the Jewish community.

During one Independence Eve of Afghanistan, my friend Yusuf came to me asking to accompany him to a musical performance at the royal theater near the palace. I asked him how we were going to get in. He responded, "I am surprised by your question since you have connections at the palace." While talking in the yard of my house, I noticed a luxurious bus with the royal insignia. I understood immediately that it was coming to pick up some of our neighbors who belonged to the royal family, to take them to the event. The bus stopped nearby and opened its door. I urged Yusuf to get on board and to sit down. The driver did not ask any questions. A few minutes later, the invited guests came and boarded the bus.

On its way, the driver made more stops and more people that we didn't know got on board, until we reached the theater. All of the passengers went inside as a group and we were among them. The whole group, including us, sat in the V.I.P. section with the royal family surrounding us. The show started with performers from Tashkent and Buchara of southern Russia. It was a great, impressive show and very enjoyable.

During the intermission, the lights in the theater turned on and my friend and I found ourselves sitting one row behind the king, the queen, and their closest family. Suddenly, one of them turned her back and asked us who we were. I still do not understand why I was not embarrassed, but I answered bravely with pride, "I am Jewish, the son of Israel Chan." She responded, "Oh, I see," and she turned back her head. I understood from her response that it was okay for us to sit there. We remained until the end of the show and then went out following the royal entourage. Part of the huge audience who had already left the theater was waiting outside to see the royal family leaving. Among them were a few friends of mine and members of the community. They asked us what we were doing with the entourage of the king. They wondered how come we were among them. I did not answer but only responded with a smile.

During the years 1963 and 1964, Afghanistan in general and the capital Kabul in particular was a real paradise on earth. All the necessary products were in abundance and the citizens were happy. At that time, we received good news from the land of Israel. The land was enjoying a period of peace and prosperity. The information of what was going on there came to us via letters, Israeli radio broadcasts, and Jewish tourists of Afghanistan who visited Israel. They had nothing but good things to say which triggered another wave of our youth to immigrate.

Friends of mine like Amnon Elimelech, Weizman Shamash, Moshe Kashi and his brother Eliyaju made *Aliyah*. The Jews call the immigration to Israel by this name since we believe that a Jew who moves to the land of Israel is achieving a personal spiritual elevation. I would not deny that the departure of these friends influenced me to want to do the same.

I approached my Dad with the suggestion to leave Afghanistan, although the conditions there were good. I explained to him that no matter what he thought, this country was still considered a Diaspora for us, and the good news from Israel was very encouraging. Mom heard this conversation and agreed with me completely. She was ready to leave even tomorrow. Dad insisted that it was not the time yet to make this move. Official agents of the state of Israel and representatives of the Jewish Agency approached members of the community with tempting proposals. They offered free lodging in Israel in case we decided to immigrate as well as financial aid to furnish the apartment, assistance in high education and job placement, and other free social benefits.

We listened to Israeli radio which broadcasted every evening in short waves half an hour of news from Israel in the Persian language. From the broadcasts, we learned that the economy there was growing fast, the Israeli army was getting stronger, and sports were advancing. The Israeli national soccer team won first place in Asia. The radio commentators urged us enthusiastically to make *Aliya* promising that we would not regret it.

Dad, who was the head of the family, and had the ultimate right to make decisions of this kind, still felt that we were not ready yet but was sure that the day would come. Dad explained to us that the reason he did not feel ready to make the move was a lack of money. He did not want to immigrate empty handed. Thus, he made more efforts to save money. His friends sent money to their relatives in Israel to buy them houses or land which were very cheap at that time, but we could not afford it. I did not foresee a substantial saving in the near future since our routine household expenses were very high.

We needed most of the monthly income to afford maintaining a comfortable lifestyle for a large family and for community necessities. Dad promised us that when he would save enough money just to provide us with supplies for the trip we would immigrate. I was sixteen years old already and eager to make something of myself. Although I was very busy with my studies and work with Dad, I found time to host friends at home especially in my little room. Usually once a week, either on Saturday night or Thursday night, a few friends came to me or we used to spend the evening singing with musical instruments.

In our yard, I built a gym complete with weights, a bench and a trampoline. My gym teacher Zykria was a real sportsman with an athletic body, and he came two to three times a week to exercise with me privately at the yard. Additionally, we often rode on bikes to nearby towns like Golbhar, Qarqa and Paqman which were at the distance of 30 to 40 kilometers from home. I was in good shape and was aware of the importance of eating healthy food to stay in shape. For example, almost every day, I drank a very tasty beverage which Mom prepared from natural goats' milk with cream above it called qymaq.

I felt self-confident and my close friends were older than me by two to four years. We went together to school parties and to weddings of Muslim friends. They prepared for us special food since they knew that we do not eat meat at their parties. Sometimes, they said to us, "We eat your food. How come the Jews do not eat ours?" We then needed to give a long lecture about how Jews are required to eat kosher food. We discussed the Torah laws regarding the separation of meat and milk products, and eating meat only of an animal that was slaughtered by a person who knows the practical religious laws of slaughtering. As they respected our belief in the Holy Torah they always responded by saying, "If it is written there, we will not dispute it."

CHAPTER 12

My younger brother Gavriel was six years old, and was a quiet but mischievous child. One morning, he left the house all alone and got on the bus that Dad used to take to work from *Shahar Naw* to *Shahar Konah*. After a long ride, the bus came to its last stop and my brother got off. He did not know where he was. At home, my mother realized that the child was missing. We looked for him all over the house and in the yard but he was nowhere to be found. We called some neighbors for help and all of us combed through the neighborhood until we reached a huge park crying "Gavriel! Gavriel!" No one answered our cries. Meanwhile, people called Dad at work to notify him of what happened. Dad immediately left work and joined the searchers with his Indian aide. This aide suggested calling the Kabul radio station to ask them to broadcast an announcement that a child fitting a certain description was missing. The family liked this idea and my brother-in-law, Tzvi, indeed called the station. He explained to them briefly what happened and asked them to broadcast an announcement: "Anyone who saw this child is kindly requested to bring him over to a well-renowned fruit & vegetable store owned by Char-Rahi Torabazchan or to call the store."

The search continued for the entire afternoon and towards evening, the congregation gathered in the synagogue for a special prayer. A short while after, a messenger came in a rush from the fruit store and told us that someone called and notified that Gavriel was found on the other side of the city and was brought to the store.

Tzvi took my bike and quickly rode to the store. He hardly talked to Gavriel but immediately seated him at the front of the bike and came back in ten minutes. It was already late at night and people who helped us with the search waited until Gavriel returned safely. The congregation was very happy upon his return, while he was shocked to see so many people giving him so much attention.

I was curious to know where he disappeared to but my parents demanded that I not ask questions and not to be angry at him, since Gavriel was exhausted. He asked for food because he did not eat for the whole day and right after, he fell asleep. I would say that this was a rare case in Kabul at that time since children did not leave home without permission from their parents. Thank G-d that it had a happy ending.

Haji Yitzchak Basal had just returned from a long tour in Israel. On many Saturdays, he used to come to our home and eat with us. This time, he strongly tried to convince us to make *Aliyah*, since it was the best time for Dad to make this move. He said, "Your children are adults now and they can work in Israel. There are plenty of jobs there now since this country's economy is growing tremendously fast." Dad did not answer but simply said, "With G-d's help, we will think about this idea only after the coming holidays. I am now busy with arranging the wedding of my daughter, Tzippora."

Tzippora had already reached the age appropriate for marriage and matchmakers from all over came with offers to Dad regarding her, but Dad refused for a long time. He told them, "We are moving to Israel soon so it is not the right time to talk about matching." At the time, I was in a company of friends who were older than me, and without my knowledge one of them had an eye for my sister. His name was Asher and he belonged to a nice and decent family who originated from the city of Balkh.

Tzippora and he had a daily, written correspondence without the knowledge of the families. One day, the rabbi of the community, *Mollah* Yosef Aharon, appeared in our house as a matchmaker. He asked Dad to approve his match between Tzippora and my friend. Dad was very surprised and as usual, refused saying that it was impossible now. The Rabbi came many times in this regard since the boy strongly insisted on this match. Then Tzippora was asked for her opinion and she gave her consent. Dad was left with no choice and understood that it was a heavenly match. He gave his consent and *shiriny* candies were given as a symbol of confirmation.

The rumors spread all over the community. It was really a great surprise for them. This groom-to-be was well known in the community as a musician who composed his own songs. Some of these songs were personally dedicated to my sister and sent to her. Even his close friends, who performed with him in his shows, were surprised. They were a music band in their twenties who performed concerts with accordion, mandolin and drums. Once in a few months, they conducted concerts for the community in the big yards of the rich members. These events always started with a sermon by the Rabbi and his blessing to the community and then the band performed a nice musical show which left the whole community happy.

CHAPTER 13

The years of the sixties were a beautiful period in the history of the Jewish community of Kabul. Members lived in prosperity, threw parties and took vacations to enjoy life. The youth made picnics in the parks and hung out in movie theaters and night clubs. Life was good, which was one of the reasons the Jews of the capital did not want to leave to emigrate to the Holy Land; despite the good news from there. The Afghan regime was in favor of the rich people of the country, of which many were Jews. The poor people of the country who were aware of this situation naturally did not like it. Demonstrations against the regime started at the universities where children of these poor people were students.

Towards the beginning of the seventies, these demonstrations intensified and were directed also against the Soviet Union who were influential in the country but did very little to advance it. In order to stop these demonstrations or at least not to let them spread out of the campuses, many policemen were sent to frighten the students by shooting in the air. Anyhow, rumors of what was going on there spread outside and reached the high schools. The impression was that a revolution against the government was on its way.

The heads of my school warned the students not to publicize these rumors. At home, Dad repeated these warnings and said that we as Jews, who are a minority, must be very cautious not to talk too much in this regard. He was afraid that in case the situation in the country deteriorated, the Jews would surely be the first group accused for this. He added that throughout history, Jews experienced this pattern in many Diaspora countries. Thus, we had to be cautious.

With time, the rumors of unrest spread all over and Jewish families started to feel threatened. Little by little, they sent their children to Israel as a safe haven. I also tried to convince Dad that our security was in jeopardy. I told him that young college graduates whom I met were realizing that the royal regime was unfair to different sections of society and they wanted it to be replaced by a more advanced and modern government. I added that the students now were better educated; they were more sophisticated, and knew what was going on in modern countries.

A few Muslim friends of mine like Yusuf, Zalmai Pahinda and Zykriya planned for the time being to leave the country to France and Russia for higher learning. They wanted to return with a valuable degree in order to achieve a good job.

That year, my brother-in-law Tzvi received a draft to the Afghan army. Usually, it was rare for a Jew to receive the draft. Unfortunately, he received it since he moved from the city of Hirat and registered in the city of Kabul, thus receiving a *Taskara* which lead afterwards to his draft. He made several times to release himself from the draft since he was married already, but did not succeed.

Bribery was very common in Afghanistan, and you could get almost anything you wanted with it. Tzvi was recruited to the army but he bribed his officer in charge with money and a few boxes of food. His plan to dodge the army succeeded. He hired an unemployed person and gave him a monthly salary to go everyday to the army base and say *Hazir Zahib* when the name Tzvi Basal was called, and right after, he could return home with the consent of the bribed officer. This lasted for a year until Tzvi received his final discharge.

My eldest sister, who lived in Israel, gave birth to a girl, my parents' first grandchild. She was named Dorit. Also, my sister Bracha, who got married a year and a half earlier with Chalifa Tzvi, was pregnant in her ninth month. The duty given to me was to call the midwife when the time came. I was shown in advance where this midwife lived. One evening towards darkness, in the middle of a very snowy winter, my brother-in-law Tzvi came to me and said,

"This is the time to call the midwife." I went to a nearby neighbor who cautiously drove me in his Volga car through narrow alleys to her house. It took her about fifteen minutes to prepare herself along with her case and we all returned home.

My sister was expecting her with a few other women in her room as it was customary at the time. Tzvi and I stood behind the door. I was stressed and was hoping there would be no problems, while Tzvi was reciting Psalms for the success of the birth. We heard neither yelling nor screaming. Suddenly, the door opened and we were called in to be notified of the birth of a cute and beautiful baby girl. Tzvi immediately thanked the midwife and asked her how the mother was doing.

On the upcoming Shabbat a *tochom chori* was thrown to honor the baby's father and to give her a name. She was named Dahlia. In our community, it was customary that if the firstborn was a girl, the kids would wait for her father to leave the synagogue to call after him jokingly, "*Maror Maror.*" Perhaps it wasn't very nice, but it was a lot of fun.

My parents were very happy to have this grandchild since she was the first in the family. My mother, with her knowledge and experience, helped the young couple with a lot of guidance and care for the newborn. A few months passed and the scheduled wedding of my sister Tzippora with Asher got closer. The whole community was invited as well as businessmen whom Dad was in contact with. The big yard at our house was prepared to accommodate this event. Tzvi helped us with arranging lighting and decorations.

The night before the wedding, as customary, the bride's hair was dyed to a reddish color at her *Hanna Bandon*. She was also taken to the *mikveh* accompanied by drums and dances. The whole week before the wedding, the couple did not see each other and they fasted on the day of their wedding.

Before sundown, all the guests had arrived and the Rabbi in charge of the wedding, *Mollah* Yosef Aharon, arranged a *Tallit* held at the edges by four people. The couple and the Rabbi stood beneath it. Surrounding the couple were little girls with lit candles in their hands. The Rabbi blessed the couple over a cup of wine and read the religious contract of marriage. As he finished reading, they were considered husband and wife and the whole audience shouted "*Mazal Tov! Mazal Tov!*"

The new couple entered a separate room where they were alone and broke their fasts. All the guests found their seats around the table on the lawn and ate appetizers until the couple came out of their room. They joined the family table to the tune of drums and then the main meal was served.

The guests sang special songs for the groom and the bride, as did the colleagues of the groom, accompanied by accordion, mandolin, and drums. The celebration beautifully repeated itself for seven nights as was customary, and the couple was blessed with *Sheva Brachot*.

The last night Tzippi was going to move out of our house and go with her husband to their new home. This was a difficult separation for our family, which involved many tears. I still remember how she embraced me tightly while sobbing, since we lived in the same room for many years. We felt sad when she left, even though her new residence was not far from ours.

In the beginning of the spring, about a month before *Purim*, the community Rabbi, *Mollah* Yosef Aharon, got sick. I was chosen on a very early morning to go to the doctor and call him urgently. The physician's house was still dark which meant that he was still sleeping. I knocked on the door and the response was, "Come in. The door is open." He was still in bed and I apologized for waking him up. I explained to him that the Rabbi is sick and asked him to come as quickly as possible. The doctor responded positively. He jumped out of his bed, washed his face, dressed himself up and took his medical bag with him. All of this took just a few minutes and both of us left his home and went to the rabbi's house.

Right upon arrival, the doctor checked the Rabbi and said he needed urgent hospitalization. "The rabbi's condition does not look good but I cannot diagnose him accurately since X-rays and blood tests need to be done," the doctor added. The Rabbi refused to leave his home, demanding the doctor to give him a prescription and saying, "I am sure everything will be alright." The doctor insisted that the Rabbi must be hospitalized and referred him to the new hospital called 100 Bistara, which had just been recently built by the Czechs. I encouraged the Rabbi to listen to the doctor since he had nothing to lose.

I told him, "We will be back soon after your exams are done." I called a car service which came immediately and we left swiftly for the hospital. After a short exam, the physician ordered immediate hospitalization. Surprisingly, the Rabbi refused to stay. I did not know what to do and called one of the community leaders for advice. In a short while, two community members, close friends of the Rabbi appeared at the hospital and tried to convince the Rabbi to stay. During their talk, the Rabbi became anxious and lost consciousness. The doctors gave him first aid and it was no longer necessary to convince him to stay.

In the coming days, the community leaders visited him daily. I also came after a few days, exactly one week before *Purim* to visit the Rabbi. I hardly recognized him because he was pale and had lost half of his weight. He spoke very fearfully and said that he did not see any chance to get out of the hospital alive. I knew that he had already notified the community to be prepared for his death and asked them to properly observe the upcoming holidays of *Purim* and *Pesach*. During my visit, he suddenly asked me if the community had rented the mill to prepare matzos for *Pesach*. I did not know what to answer and I returned home. The day after, my Dad went to visit the Rabbi and very soon came back home. He said that the Rabbi was not functioning anymore and only a *Purim* miracle could save him.

Unfortunately, a miracle did not occur and the Rabbi passed away on the early morning of *Purim*. His body was brought to the yard in front of his house and the Rabbanit embraced him while screaming and crying bitterly. It was necessary to use some force to separate them. She then instructed the community leaders to transfer her husband's body to be buried in Hirat since his whole family lived there. The local Jewish *Chevra Kadisha* took care of treating the rabbi's body in the traditional way and a bus was rented to move it together with the accompanying people from the capital in route to Hirat, a twenty-four hour drive. My new brother-in-law decided to be among those people out of gratitude since this Rabbi was his matchmaker and just recently married him. Deep sorrow and grief befell the Jews in Kabul and all scheduled celebrations for the *Purim* festivities were canceled.

Since the *Pesach* holiday was very near and needed extraordinary preparations, an acting rabbi was appointed immediately for the community. *Mollah* David Moradoff was chosen for this position. He replaced the late rabbi also as a *shochet*. He was young, smart, pleasant and beloved by all. Besides being a qualified cantor, he was thoroughly familiar with a few languages and among the very few in the community who was fluent in English. I remember how he introduced our community in English to representatives of *Chabad,* a huge Chassidic community in Brooklyn, NY, who came from America to visit us. Rabbi Shamuel Pesach Bogomilsky, and Rabbi Kopol Bacher. They gave us strength and very good feeling…

CHAPTER 14

The Jewish families who lived in *Shahar Konah* moved to the other part of the city where we lived, in *Shahar Naw*. The congregation here grew tremendously and the leaders decided that it was time to buy a big lot in order to build a new Jewish community center. Fortunately, the lot that was chosen was not far from our home. The leaders, *Mollah* Shalom Shamash, Shmuel Aharon, and Levi Haji were in charge of the construction and did the best the could to finish it as fast as possible. *Mollah* Shalom Shamash used to come in the early morning to supervise the laborers. The plan of the building included on its first floor a *mikveh*, a children's *Yeshiva*, and a residency for my sister Bracha and her husband Tzvi who would manage the *Yeshiva*. The second floor would include two synagogues – a big one for the whole congregation and a small one for midweek prayer for the youths.

Bracha and Tzvi were very excited that they would soon move to a brand new place especially since my sister was eight months pregnant. They were eager to have a son and reserved the honorable position of *Sandak*, in case it happens, for our father. In her ninth month, the building was ready and the synagogue moved to its new location. My sister and her husband got a three-room apartment and an extra room on the second floor assigned for guests who came from outside the city. The whole building was modern and included an electric refrigerator, which was not yet found in any other home. Even a private telephone was installed in their apartment.

One of the first phone calls was to the midwife to stand on alert in the coming days. Two weeks before *Pesach*, my sister felt ready to give birth. Some members of the family went to the midwife's home but unfortunately, she was missing – she had gone to a wedding. They returned empty handed and deliberated whether to take her to a hospital. Suddenly, someone of the community recalled that he knew an old, retired midwife and maybe she would be ready to come and help. Again, we left to her home and after a lot of persuasion and promises to pay her well she agreed to come. My sister 'welcomed' her with yells and screams and the experienced midwife helped her to give birth right away.

What a joy it was to see a baby boy joining the family. This boy came into a nice era. Spring had just begun and the community moved into a new Jewish center. The Brit of the newborn took place on *Shabbat Hagadol* in 1966 and it was the first one to be celebrated in the new synagogue. Dad took care of all the financial expenses for the celebration meal and did all he could to make sure that the many guests would be satisfied. Dad was very excited to be the *Sandak* and the boy was named Yosef, after the late Rabbi.

For *Pesach*, our whole family was guests at my sister's home since she and her husband did not want us to feel lonely on the holiday. Our big house, which was always crowded with people since we shared the same yard with the synagogue and the Jewish community center for over ten years, suddenly became quiet. The whole Jewish complex moved away to its new location.

We did not need such a big house anymore, but the huge expenses to maintain it remained as before. For all these years, my parents did the utmost in helping the synagogue and community center to function in the best way. They did all this willingly with much joy and as they said, "It is all for G-d's sake."

Winters and summers, nights and days, Saturdays and holidays, they were always busy. They never complained in this regard. Some people of the community appreciated their help and thanked them for all the efforts they invested during so many years. However, there were also people that complained, but they did not do so enthusiastically.

When the synagogue moved to its new building, the *Gabbaim*

distributed seats for the members according to their importance in the community. Thus, they determined the seat of my father to be near the *Aron Kodesh* saying that he deserved it as he served the congregation faithfully all the years before. However, Dad had already picked up a seat for himself at the extreme opposite location, the first seat near the front entrance of the synagogue. The Gabbaim tried to convince him to move to the seat which they determined for him, but all of their requests did not help. Dad said that the location of his seat in the synagogue does not show his importance. He added, "The importance of a person is determined by his behavior and all seats are equal in my view."

A few months passed for the new synagogue and we, noticed that Dad's mood was not as before. He was not responsible for the congregation anymore and in his business he did not succeed to become richer. My brother-in-law Tzvi and I consulted with each other, and came to the conclusion that now was the proper time to move to the Holy Land. We told Dad, "If until today you have not made a fortune, there is no chance to make it now at your age. It is too late." Many of his friends took advantage of the periods of prosperity in the country to accumulate a fortune but Dad was too busy in spiritual tasks which brought him much happiness.

On one Shabbat, my sister and brother-in-law were invited to us for dinner. This was the time that Tzvi decided to put full pressure on Dad to decide to move to the Holy Land. Dad could not contradict our reasonable arguments that that time was the best to make the move. He raised a practical argument that he does not think that the state will issue passports to go abroad for such a big family as ours. Tzvi did not give up and said, "Leave this issue to me. What I want from you is just to get the OK for the move. All the rest leave for Michael and I to take care of." Surprisingly, Dad gave his consent.

Hearing this, Mom became extremely happy since she wanted to go to the land of Israel for many years. Tzvi and I didn't speak with my Dad on the subject again, but we started to move things forwards in a fast pace. A photographer was invited to take passport pictures of all family members and we went from one office to another to get the required approvals. In each place, we were asked the same questions: Where are you going to and why? Our answers were always that we were going to Iran to attend a family wedding. This was a plain idca which seemed to work very well. Every day, we returned home with good news regarding our advancement but Dad was still pessimistic of our success.

My Father was under terrible financial pressure since all the expenses of the huge house and yard was solely on his shoulders. As far as I remember, it came out to 3,000 *rupee* per month which was a huge amount at that time. One day, I informed Dad that *Mollah* Nissan Yazdi, one of the prominent members of the community, had three vacant rooms which would have been quite good for our family. I asked him, "What is your opinion of offering him 500 dollars per month and move there?" Dad liked the idea and on the same day, he sent me to talk with Nissan. He welcomed me very cordially and immediately agreed to the offer but added, "I do not want any money from you, and you may make the move right away." I returned home immediately and told Dad the good news.

We settled all the matters with the landlord and threw out all the junk which had accumulated in our big house over the course of over ten years. Within one week we moved to our new home with all of our necessities. This place was just two blocks away and contained only two rooms, one big, long room for my parents and all the little children, and behind the building was a small room in which only two beds were placed. One bed was for me and the other was for my cousin, Chai Cohen of Hirat, who used to come very often for business in the capital. My Dad's monthly expenses dropped tremendously and he became more relaxed.

The year was 1966 when Tzvi came to us one day with passports and visas that were valid for one year. Although we had time to decide if and when to leave the country or not, Dad did not change his mind and immediately determined the time of departure, right after the *Pesach* of 1967. The itinerary was planned amongst the family to go through the city of Hirat to Iran and there, the Jewish Agency would take care of the rest of the move.

Excitement grew within the family from day to day, especially for my mother, since all of her parents and siblings, whom she had not seen for almost thirty years, lived in the Holy Land. I helped Dad to liquidate all of his businesses, collect debts, and buy items that were expensive in Israel. I bought a few carpets and electronics which I intended to sell in Israel besides keeping for our own use. Dad also asked me to buy items which I bought for myself for my brother Amnon, whom we had not seen for the last twelve years.

At school, I finished the tenth grade and had just started the eleventh. My friends were already informed that this was my last year in school since I was leaving to Iran. Only my close friend, Yusuf, knew the truth that we were going to Israel.

Throughout this time, we received encouraging letters from our relatives in Israel who were impatiently waiting for our arrival. In their letters, they also wrote detailed advice on what to bring along with us such as necessities for a house, referral letters from school, diplomas, drivers licenses, and more.

The winter of 1966 arrived and we stayed home a lot since as usual, the weather outside was snowy. We hung out with our neighbors in the building, *Mollah* Nissan Yazdi and his wife, Tamar, as well as with the widow of Rabbi Yosef Aharon who lived at the front of the building. The two sons of the Yazdi family, Amnon and Elimelech, immigrated to Israel two years ago and always wrote encouraging letters from the Holy Land. Whenever the Yazdi's saw me, I reminded them of their sons whom they missed very much. The Rabbanit always asked me to visit her home in the evenings in order to teach her nieces English. I fulfilled her request frequently until she moved to the city of Hirat upon the invitation of her uncles.

Our time to leave the country was approaching. The *Hanukah* festivities of the winter passed and *Purim* and spring were approaching. A week before *Purim*, we received an invitation to be guests of a tribe called Saaid Kayhan. This tribe was in good relations with my father and other families of the community. They liked Jews and always used to send us gifts. They also visited us from time to time.

Dad received the invitation personally for our immediate and extended family, which also included our brothers-in-law Tzvi and Asher and our guest from Hirat, Chai Cohen. Another family also got such an invitation was Levi Haji's who was joined by their guest from Hirat, Eliyaju Basal. All of us went together in three cars to the mountains where the Kayhan tribe resided. It took us about an hour drive from Kabul on narrow and unpaved routes to reach the place. The entire way, we were ascending mountains seeing beautiful views and driving very slowly. When we entered the village, people of the tribe were standing on both sides of the road to welcome us. The leaders of the tribe accompanied us to a building which looked like a palace and was located on the edge of a mountain. They gave us a few rooms which were luxurious and modern, the likes of which I had never seen in Kabul.

We placed our belongings in the rooms and were asked by our hosts to come down to the tea room after taking a rest from the ride. At the tea room, we conversed and it was obvious that the host did the utmost to make us feel comfortable. They planned for us a schedule for the three days of our visit. Since we ate only kosher food, my brother-in-law, Tzvi, brought with him a special slaughtering knife which he used every day to slaughter chickens and turkeys for our meals. In the evening, we ate dinner only after taking a drink of whiskey together with our hosts. They notified us that the morning after, we would go horse-back riding and afterwards, there would be a *bozkashi* game that would have been a shame if missed.

After our usual morning prayer, we ate breakfast and then went to the stable. The air on the mountains was clear and fresh and we felt great after receiving a horse for each one of us. I was in the group of my two sisters and brothers-in-law and the last one to receive a horse. We ascended the horses and the group started to move without me. I was shouting, "Wait for me! Wait for me!" and they looked back to see that my horse was not moving. My horse looked very tired and was blind in one eye. They started to laugh at me because of my misfortune with the horse. My brother-in-law, Asher, shouted "Hit him with the whip!" I hit the horse and he started to move, but very slowly since he was tired. They quickly rode to reach the *bozkashi* field. Of course, I was the last one to arrive there and they laughed at me again.

I asked my sister, Bracha, to ascend her horse to pose for a picture since her horse was the best in my view; beautiful and white with a noble demeanor. Besides posing for a picture, I also wanted to have a good riding experience on such a horse. Suddenly, a horseman dressed in the uniform of a *bozkashi* player approached us quickly on his horse. To our disbelief, we realized that this horseman was our Father. "Are you playing?" We asked him. "No," he answered. "I was only honored to open the upcoming game."

The horsemen arrived at the field and all the men of the village were gathered around. The women returned to the building where we were stayed to prepare lunch. The players were sitting alert on their horses surrounding the circle in the center. At the first signal to start the game, the horsemen jumped forwards to catch the goat, spurring the horses with whips. This was a tough and cruel game. We kept distance and stood on a hill since the playing field was an open one and any horse could have galloped out and hit the spectators. Suddenly, I saw on horseman snatch the goat and gallop quickly towards the goal. In one hand, he was holding the goat and with the other hand, he was guiding his horse. On his way, players of the opposing team tried to snatch the goat away from him. However, he succeeded to maneuver away from them and win. To have that knowledge and experience, the horsemen are trained intensively and strong, fast horses are assigned to them.

At noon, we returned to the building and a great Afghan meal of *ash palow* was waiting for us together with turkey and vegetable salad in tomato sauce. We were served like kings, which I will not forget. We would have liked to stay there for a week but we had to return after another two days to our community in Kabul due to the upcoming *Fast of Esther* and *Purim*. We spent these days exploring the surroundings and playing games such as *Toop Danda*.

On the last day, the tribe supplied us with a sheep to be slaughtered but my brother in law, Tzvi, could not do it since he was ordained to only slaughter chickens and turkeys. We therefore ate turkey. The separation from the tribe was very emotional and we thanked them very much for their wonderful hospitality. They actually wanted us to stay longer but we explained to them that tomorrow evening would be the start of the *Purim* festivities and we needed to attend the special evening prayer in the synagogue which includes the reciting of *Megillat Esther*.

The next day, we woke up in the early morning before sunrise and ate the last meal before the fast of that day. We left the tribe early in the morning so we could reach our homes by noon time for *Purim* preparations. This time, the ride was easier since we were going down hills but the path was still crooked and unpaved. My cousin, Chai Cohen, was driving the first car when suddenly, he hit a stone and the oil tank was damaged. The oil started to leak out and Chai stopped the car to check the oil. There was almost nothing remaining, but to our good fortune, we succeeded to reach a nearby garage to check the tank. The mechanic immediately discovered a hole in the tank, and said that we had to wait until the tank could cool, and then he could fix it. Chai Cohen did not want to delay the whole convoy since *Purim* was getting closer. He had experience with tricks and asked for cotton and raisins. He crawled under the car and sealed the hole by gluing the cotton to patch the hole with raisins. He then filled the tank with oil and bought a few more oil cans for the journey. I was in the second car, a Land Rover, which was crowded with ten people pushed in; five adults and five children. We drove very quickly and reached home at noon as planned.

The women immediately started to bake special cookies for *Purim* such as *Ozney Haman, sambosa*, *klocha*, and *lekah*. In the evening, we went to the new synagogue for the start of *Purim* ceremony and were extraordinarily welcomed by the congregation since this was our last year in Afghanistan. They wished us the customary wish *Le'Shana Haba'ah B'Yerushalayim*, next year all of us in Jerusalem.

We noticed that they were starting to feel the differences from the old location of the synagogue, where we spoiled the congregation greatly by taking care of all of their necessities for over ten years. My parents also felt sorry since they started to feel how much their activity for the congregation was important for them. All they did in this regard was done with all their hearts and with much love. Regardless, they were very happy to see that they were leaving the congregation in an excellent condition when they just moved to a big new building.

This itself was a another good sign for my parents that it was the right time to move to the Holy Land. We still had to pass the *Pesach* holiday and right after it, we would leave. Our plan was to celebrate the *Shavuot* in Israel. In Afghanistan, we called this holiday *Mo'ed Gol*. It comes out in the beginning of summer and it was definitely good timing to start our new life in the land of Israel. We scheduled the thirty-first of May, 1967 to leave Kabul by bus in route to Hirat.

Closer to *Pesach*, we were guests at my sister and brother-in-law's apartment located in the new building complex of the Jewish community center. Since it was a brand new apartment, they did not have to do much cleaning for the upcoming Passover. Their two little children, Dahlia, who was two-and-a-half years old, and cute Yosef, who was almost a year old, were beloved grandchildren to my parents. No doubt that the upcoming separation between them would be emotionally difficult. However, the parents of the children promised to follow our immigration to Israel very shortly after.

Pesach passed and right after, many people came daily to bid farewell. Among them were also teachers and students and even Muslims who simply knocked at our doors. They liked and respected us very much. My friend Yusuf came every day to help me. He was like part of our family; a very smart guy who found interest in politics. In his view, a war was coming very soon between Israel and its Arab neighbors. We listened nightly at eight p.m. to Israeli radio broadcasted in short waves in the Persian language. The reception was very good so we were updated on everything that was going on in Israel.

The country before 1967 was very small and surrounded by tens of millions of enemies. At that time, there were often clashes between Israel and Syria, and Dad was very worried. However, we did not renew the lease for the house and all of our belongings were already packed into suitcases. Therefore, we did not have the option to change our minds but to go ahead with the plan to immigrate. We had our full trust in G-d that he would not let Israel down, besides knowing that the Israeli army was very strong.

I was already seventeen years old, very mature, and behaved like an adult. I was very familiar with what was ahead of us since I was guided well by my two brothers-in-law, Tzvi and Asher. I impressed my Father with my self-confidence and gave him the feeling that he had nothing to worry about as long as I was near him. He trusted me indeed. Four days before leaving, Tzvi and I went to the central bus station to buy tickets for the bus to Kandahar. The plan was to stay there overnight and then to take another bus to Hirat where we would stay with the Jewish community on Shabbat. Then we would go to Israel via Iran.

For the remaining few evenings, we were invited for dinners with community members as a show of gratitude for what we have done for the community. In the last evening, the whole community came to the synagogue to say goodbye to us. *Mollah* Babajan Vardi delivered an exciting speech on behalf of the community in which he told the audience that he knew our family personally for many years. He emphasized my parent's devotion to the community, while others were primarily busy in making money. Afterwards, he thanked my parents very warmly and wished us a safe trip and much success in our new country.

He finished his address expressing the hope that the whole community would move one day to the land of Israel. The whole evening was very exciting for us as well as for the community. This was my last night in Kabul; the place where I was born and was going to leave with good memories. This separation was not easy for me since the many friends whom I was leaving behind were very nice and generous; typical of this country. I grew up in Kabul in a warm and pleasant family surrounded by a supportive Jewish community who were a minority in a big country and nevertheless, never suffered any racial discrimination. I was sure on that evening that I would miss this country, which I felt much gratitude for, but as a Jew, I understood that my future was in the Jewish state.

The last night in Kabul, I couldn't fall asleep, so I turned the radio on to listen to Afghan soul music performed by Ostad Rahim Bacsh. Many times in the late hours of the night, I liked to listen to this singer with his relaxing voice. The little room where I lay on my bed was full of suitcases that were already packed for the trip. Towards morning, I managed to get some sleep because I was extremely tired, but then the roosters' call woke me up for the morning prayer. I said to myself, "Today we are leaving for a long journey in which I am leading the family. This is not a time to feel tired." I immediately got out of bed. Dad had already awoken the others. Tzvi opened the synagogue for the early morning prayer and my sister Bracha, the owner of the apartment, prepared breakfast as well as food for the journey. Outside, the chauffeurs were already loading the suitcases and belongings into the cars. My two brothers-in-law, Tzvi and Asher, and my sisters, Bracha and Tzippora, decided to accompany us until the city of Hirat which was a far, eighteen-hour drive. Dad was under heavy stress and we tried to calm him down. My mother, on the other hand, was very relaxed. Tzvi and Asher continued to give me instructions for this long journey until the last minute, which they did very pleasantly.

We finished our breakfast quickly, went out to the cars that were waiting for us, and drove to the bus station. The bus was to take us first to the city of Kandahar which was a ten-hour drive. At the bus station, we saw a few people and I thought that maybe I would recognize some people that had not yet said goodbye to us. To my surprise, I saw my friend Yusuf who had come especially by himself to say goodbye at the last minute. The porters took the suitcases onboard the bus and the whole family got on and took their seats. I sat in the first seat of the right-hand side of the bus, while the rest of the family took their seats at the left side. At that point, my father was more relaxed. Most of the bus riders were tourists from America and England who had come to see this beautiful country especially in the spring season.

When the bus driver announced that in a few minutes we were leaving, I noticed that Yusuf was still standing outside beneath a tree, observing the bus with a sorrowful look on his face. I could sympathize with his feelings because I felt the same. It was a separation between two good friends who had been together almost every day for many years. I got off the bus to embrace him and give him a last kiss. He was very moved emotionally. I heard the driver start the engine and got back on the bus. All of the family seated recited *Tefillat Haderech* and the bus started to move. We traveled parallel to the Kabul River, which was very turbulent because of the melting snow that flowed into it from the northern mountains. We crossed the old city of Kabul and then through a large desert that had just a few villages of Bedouins living in tents surrounded by flocks of sheep and goats.

Our family on the bus comprised of fifteen people; seven adults and eight children. We made up the majority of the passengers on the bus. The rest were tourists. As it was a long ride, Mom took out a plate of dried fruit and cookies which she served to everyone on the bus, including the driver. The driver was thankful and ate what was given to him while driving very fast on the highway. This highway, which connected the cities of Kabul and Kandahar, had recently been paved by Americans. Some of the passengers took naps including my little brothers, Gavriel, Rafael, and the youngest Avraham, who was then five years old. I was busy for most of the ride studying to improve my English from a Persian-English dictionary. Another purpose for holding this dictionary in my hands was to answer questions that the tourists might have asked me.

Around noontime, the bus arrived at an inn for a lunch break of one hour. Since the weather there was warmer and drier than in Kabul, my brothers and I jumped into the swimming pool of the inn to refresh ourselves. Some of the male tourists joined us. We got out after ten minutes, and an American tourist approached me asking for a word to be translated. I did not know the word until he made signs from which I understood that he was looking for a towel. I helped him out and apologized for my poor English. Mom gathered the whole family for lunch under a huge tree. She gave us *yachny* with bread which she had prepared back home. Just as we finished eating, the driver announced that we had to continue the journey. All of us got on the bus and we reached the city of Kandahar towards evening.

We reached an inn in the center of Kandahar and all of the passengers were concentrated into one big hall to spend the night. We bought there a ticket for the morning bus which would take us to Hirat. Before going to sleep, exhausted from the long trip that we took, we sent a telegram to our cousin Chai Cohen of Hirat notifying him that we would reach his city the next day in the evening. We ate a bit and fell asleep. My two brothers-in-law woke up in the early morning and got out to the street to buy some food for our breakfast. They returned with fresh bread, a very tasty cheese called *qaymaq*, and pitcher of tea. We ate in the hall and went out to the bus, which was waiting at the front of the building. Nevertheless, before boarding the bus, Tzvi could not resist running to buy a few bags of fruit since Kandahar is known for its delicious fruits. The bus crossed the city of Kandahar and drove fast on the highway to Hirat. The road was not as smooth as before and it took us eight hours to reach the edge of Hirat.

We were expecting to meet Chai Cohen at the central bus station of the city, according to the telegram we mailed him the day before. To our surprise, we saw him with his American jeep parked at the side of an intersection near the entrance of the city. When the bus made a stop at the intersection, Chai approached it quickly, knocked at the door to get on, and requested the driver to follow his jeep. He told him that he was our relative and wanted us to go directly to his home with all our suitcases and belongings instead of being dropped off at the bus station. The driver, who was a very nice guy, immediately agreed and Chai spent another minute on the bus to say hello to us before returning to his jeep. The bus followed the jeep and reached the Jewish neighborhood of the city called Bazar Iraq, where he came to a stop at the front of my cousin's house. The driver and his assistant unloaded our suitcases and moved them into the house. We thanked them and the bus continued to its final stop.

Our cousin's house was huge and surrounded by a very large yard. First, Chai showed us our rooms. Asher, my sister Tzippora, and I were given the attic which was a nice room but had no beds. Mattresses and blankets were placed on Afghan carpets. That was our place to sleep. My parents and little brothers received a big room on the ground floor. We then gathered together in the living room where Chai welcomed us very nicely. It was covered with beautiful carpets and on them were placed cushions to sit on as was custom in the Oriental world. This was in the month of June when the weather was hot. While sitting in the living room, we cooled ourselves with *pakah* while flies buzzed around us. Chai swatted at them with a towel but they returned after a few minutes. It seemed that they liked us.

We were served a hot dinner while some of the Jewish community members came to see us. My parents and my brother-in-law Tzvi, who were born and grew up in Hirat, eagerly conversed with the visitors asking them about people and places which they left behind many years ago. The time was getting late and Chai stopped the conversation urging us to go to sleep, since tomorrow would be Friday. Then we would see the whole community in the big synagogue where my parents used to pray twenty-seven years ago. We were very tired from the day and immediately went to bed. Interestingly, because of the heat, the young members of Chai's family slept outside in the yard and on the roof top.

CHAPTER 15

We woke up in the early morning for the prayer in the synagogue. The city of Hirat was not the same as it was many years before. Only twelve Jewish families had remained in the city, led by the wealthy *Mollah* Reuven Amballu. When he saw us, he welcomed us very warmly and immediately invited us to his home. Since we were supposed to stay in the city for one week, Dad promised him a visit for one evening or even a full day at his house. After the morning prayer and breakfast, my two brothers-in-law and I went out to tour the Jewish neighborhood as well as the main market. We also met there tourists from Europe and America. The city was totally different from Kabul. The modernization had not yet arrived here and everything was old fashioned. The houses were old and made from asphalt and people were dressed differently. We could see the poverty everywhere and the tourists found interest in the difference between what they saw here and their developed countries. However, they realized the good character of the Afghan people who accepted foreigner with open arms.

After a short tour, we returned to Chai's house to prepare for the upcoming Shabbat. My cousin Chai was busy buying meat, fish, and other delicious food for the Shabbat and the women were cooking in the kitchen. Tzvi went out again to the central bus station and returned with tickets for the ride to Iran via Mash'had on the coming Wednesday, the seventh of June. I had a transistor radio with me so I could listen to what was going on in the world, especially in Israel. These days there was high tension between Israel and its Arab neighbors. Egypt blocked access from ships entering the southern seaport of Israel. This was a blatant act of violence since that harbor in the town of Eilat was the only Israeli connection with the Far East via the Red Sea. All Israeli imports and exports to the Far East, Australia, and Japan went through this route.

Egypt made a coalition with Jordan and Syria threatening to attack Israel in case the state tried to clear this blockade by force. Israel was a tiny country, surrounded by hundreds of millions of hostile Arabs, but it had a very powerful army. Even though the army was small, it was well armed with the most sophisticated weapons, such as modern combat planes from France, tanks, and artillery from the U.S. Nevertheless, the state needed a lot of luck to prevail in this crisis.

The security council of the U.N. warned both sides not to be the first to open fire. Tremendous stress befell the people of Israel which led to unification between opposing political parties. The leading opposition party, the Likud, headed by Menachem Begin, joined in a coalition with the ruling Labor party headed by the Prime Minister Levi Eshkol. These two leaders appointed General Moshe Dayan as defense minister. This ad-hoc appointment was part of a psychological war inflicted on the Arabs since they were afraid of him. He was a courageous warrior who lost one of his eyes in the 1948 War of Independence. Our family on the route to Israel was totally calm regarding the fate of the state since, as always, we trusted in G-d as well as in the unity of the people there. We were really impatient to get there.

Our host, Chai Cohen, asked us to go one by one to the *hamam* he had in the house so we would be prepared for the coming Shabbat. In the evening, all of us went to the synagogue and were warmly welcomed by the small congregation. According to our plan, we would be in the land of Israel by the next Shabbat if no war would break out there in the meantime. Otherwise, we would be the guests of the Jewish Agency in the capital of Iran, Tehran. Naturally, we prayed that the crisis would end peacefully. On Saturday evening, some youngsters whom I met in the synagogue in the morning came to me with the offer to go out for a walk in the city. We hung out in the main street which was dark at night. Only the lights from the houses were illuminating the street. Suddenly we saw a poster notifying that the famous singer, Ham Ahang of Kabul, was performing in one of the city clubs which was about a fifteen minute walk from where we were. I told my new friends that I knew this singer personally since he used to perform very often at Jewish parties in Kabul.

We went to this club which was at the edge of the city in a dark street. At the entrance, the doorman stopped us and asked what we were doing here. I told him that my name was Michael and that I wanted to meet the singer and say hello. He entered the club and returned right away to invite us in. The doorman said, "The singer has not performed yet and he wants to see you. Please follow me." We went in to a room backstage and saw the singer and his staff sitting on the floor which was covered with carpets and cushions. When they saw us, they rose to their feet and Ham Ahang approached me with a hug and a handshake. He asked me what I was doing in the city of Hirat and I told him that my whole family was here en-route to a visit in Iran. We sat there for about fifteen minutes and talked about the adventures we had in the Jewish community in Hirat while he was performing in our parties. He was very admired by all of the Jews in Kabul for his tremendous voice with Afghan soul songs. We separated and my friends who accompanied me were very excited from that meeting.

The next day, my family had meetings with Jewish businessmen and afterwards went to the Jewish cemetery where my grandparents are buried. The late rabbi of our community in Kabul, who passed away last *Purim*, was also buried there and his wife accompanied us on this visit. In the evening, we were invited for dinner over at *Mollah* Aghajan Basal, the father of my brother-in-law Tzvi. A few leaders of the community were also invited to that dinner. They welcomed us very warmly since in three days we were going to leave Afghanistan to go to Iran. After a delicious meal, we went out to a porch where we sat down and listened to a big radio which *Mollah* Reuben Amballu, the leader of the community, brought with him for the special purpose to listen to what was going on with the Israeli-Arab conflict. He asked us to be silent so he would not miss any word.

The tension between Israel and its neighbors was at its peak. It was after three weeks of the highest alert in Israel, which was forced into a total mobilization of its manpower. Its main supplies of vital imports from the Far East via cargo ships were blocked by a belligerent act; the ships were forbidden by the Egyptians to pass through the Suez Canal. Israel was facing the greatest peril to its existence that it has known since the hour of its birth. Multitudes of people throughout the world trembled for Israel's fate. For the first time, we heard the voice of Moshe Dayan on the radio. He spoke in Hebrew which we did not understand but Reuben Amballu translated it to us. The Egyptian president, Jamal Abdul Natzer, threatened to throw all the Israelis into the Mediterranean in case they dared to counter-attack the Arabs. We, the tiny community in Hirat, were naturally very worried about the situation but trusted in G-d, and all we could do was to pray for Israel's victory. Unbelievably my family was not deterred from sticking to its plan to go and settle in the land Israel in this critical time.

On Monday morning, the fifth of June 1967, my two brothers-in-law, Tzvi, Asher, and I went out for a tour of the ancient city of Hirat. We passed parks and mosques which were built hundreds of years ago and took photos of the three *menar* as well as of other beautiful sites throughout our exciting tour. On the way back, around noontime, we passed the market, *Bazaar* Iraq, and met there Tzvi's father, *Mollah* Akajan. He was pale with panic and urged us to return home immediately. He told us that he had just heard over the radio that a total war had broken out between Israel and its Arab neighbors. The choice for Israel was to leave or perish.

We immediately returned home at met my parents in a terrible panic. Mom said, "What will happen to us? What should we do now as we have no home? We are sitting on our suitcases in the houses of relatives in route to a zone of war!" Dad was very angry and said, "What did we do? What was missing in our luxurious house in Kabul that we made this irresponsible move? What should we do now that we have tickets for Wednesday to continue our journey towards the holy land which is in war?" The people of the Jewish community tried to calm us down saying, "You should not personally worry since we are not strangers and you are in good hands. Stay with us. The holiday of *Shavuot* is very close and by then, all the difficulties will be behind you. It means G-d will definitely make a miracle and the war will be over." Dad responded "*Tvakal Bachoda.*"

In the evening, we were invited to the head of the community, *Mollah* Reuben Amballu's home together with other members of the community. The weather was very pleasant so we all sat outside in the garden on cushions. Reuben was holding his big radio again and all of us were waiting to listen to the evening Israeli radio broadcast in Persian. In our view, this station was the only one in the Middle East which brought true news to the listeners from the battlefields. All of the other stations broadcasted fabricated facts as part of a psychological war against Israel. Until the scheduled time of the Israeli news, we were tuned to the Afghan radio which reported that the Egyptians were defeating the Israeli army and were getting close to Tel-Aviv. We really started to panic.

Suddenly, Reuben changed the channel as the Israeli news broadcast began. The perception we got then was totally different. The reporter did not get too much into details but for us it was enough to relax as he said that the Israeli air force had already succeeded to destroy all the military airports of Egypt, Syria, and Jordan. Thus, all of their air force was put out of action. All of this happened on the first day of war.

Reuben Amballu turned off the radio and in an aggressive speech to the guests he said, "As your leader, I am demanding that no Jew will leave his house or the synagogue unless it is very necessary. It is not the time to show ourselves in front of the Muslims when the Israeli army is defeating their brothers. I hope that G-d will help us and the war will be over soon." One of the arrogant youth said, "I don't think that you are correct. We cannot know how long this war will continue and you suggest sitting at home? We will explode from boredom." Reuben got nervous from this remark. He took off one of his shoes and chased the youngster yelling, "You impudent young man! Aren't you ashamed of your criticism? Get out if you want and afterwards, don't come asking for our help." He then threw the shoe at him and returned to his seat. All of us were scared from this weird sight.

Afterwards, everyone calmed down and Reuben said to my parents, "You don't have to worry. You are in good hands as long as you are with us. You are part of the Jewish community here and are warmly invited to celebrate the *Shavuot* holiday together with the whole community of Hirat. G-d bless Israel."

On Tuesday morning, my brother-in-law Tzvi took a bike and speedily rode to the central bus station to return our tickets that we bought for Wednesday. He went into a travel agency and bought airplane tickets for himself, for my other brother-in-law, and for their families to return to Kabul. They had to leave Hirat by Wednesday since their work was waiting for them, particularly for Tzvi who was a teacher for the children in the *Yeshiva* as well as being in charge of the synagogue. Asher was employed by a businessman who could not get along without him. It was clear to us that they would celebrate *Shavuot* at their homes in Kabul while we would do the same in Hirat.

The following days of that week were more or less the same. The war continued in full force while we were waiting for the evenings to listen to the Israeli radio news. All of the other Middle-Eastern stations continued with their false reports of great successes to the Arab armies. We did not go outside during those days but remained home very worried. It was difficult for us to understand how such a tiny state could cope with so many enemies.

During the nights, we could hardly sleep since we thought that the state of Israel was, G-d forbid, on the verge of destruction. At least we were among relatives and friends in the city of Hirat and not wandering to Iran as planned. We found much comfort in trusting G-d. In the coming days, we learned via the Israeli radio that in the first hours of war, the Israeli air force destroyed most of the neighboring

Arab countries' combat airplanes in a sneak attack while they were still on the ground. Then the way was open for Israel's forces, under General Rabin's command, to smash their way to the Suez Canal, to the entire length of the Jordan River, and to the Golan Heights, a strategic area which enabled Syria to overlook the whole northern part of the land of Israel since its establishment.

On the Wednesday we were supposed to leave, my brothers-in-law sent telegrams to a person in Mash'had and another in Tehran notifying them that we would not arrive on time as planned. We would stay in Hirat for the *Shavuot* holiday and possibly two weeks later, we would continue our journey. In any case, we would have let them know the exact time of our arrival.

The weather those days was very pleasant and we used to spend many hours sitting in the yard discussing the situation. Among the Muslim majority in the city there were no demonstrations or propaganda since they thought that the Arabs were in the process of winning the war. Within three days of the war, Israel conquered the whole Sinai Peninsula from Egypt, an area which was much bigger than the whole land of Israel, and the Suez Canal, which connected the Red Sea to the Mediterranean, was under Israeli control.

The security council of the United Nations demanded both parties to stop fighting, but the Israelis pretended not to hear this, since they were busy in a war of survival. The real truth was that the Israeli army was in a momentum of advancement thus the political leaders did not want to stop until regaining the holiest sites of the Jewish nation. Jerusalem was reunited and the Western Wall was restored to the Jewish people after two decades of separation inflicted by Jordan, which had invaded the city during the war of independence in 1948. Hebron, the ancient city of Jewish history with the graves of our patriarchs Abraham, Isaac, and Jacob was again in Jewish hands. The whole war was over in six days, a miracle that stunned the entire world.

The community of Hirat was overwhelmed with happiness and could not believe what had just happened. Dad's repeated response was "*Choda Bozerge*. The dreams of the Jewish people for so many generations came true and we are in the best time to move to Israel."

On the morning after the war was over, my cousin Chai Cohen and I, went together very proudly to the central bus station to acquire tickets for the bus leaving Hirat to Iran on the Wednesday after *Shavuot*, the twenty-first of June 1967. On Tuesday evening, the last night we spent in Afghanistan, the whole community gathered together to bid us farewell. This was a wonderful community of real, decent people who did only good to others and helped us tremendously

Above: Wedding of my oldest Sister Esther and Levi Simantov in Israel. They were a great help to us when the entire family moved in 1967.

Above and Right: Wedding of my sister Bracha to Chalifeh Tzvi. They became leaders in keeping the religious aspects of the community intact.

Above: Wedding of my sister Tzipora and Asher Levi. Picture with myself at the night of the wedding. Kabul (1964)

Right: Wedding of my sister Tamar to Shimon Cohen in Israel (1969). Later, Tamar and Shimon moved to the United States and were a tremendous help for me when I arrived in 1974.

Opening of a Community Center in Kabul, March 1965. Honorary ribbon cutting was given to one of the community leaders Mullah Shalom Shamash. To his left is his in-law Mullah Shimshon Kashi. They were instrumental with the opening of the new Community Center.

Left: City of Hirat June 1967. On the way to Israel my family was delayed there for two weeks due to the Six Day War. We were welcomed warmly by the community's leaders. City of Hirat was beautiful and we enjoyed the site-seeing.

Below: My brother-in-law Tzvi and I in the city of Hirat on June 5, 1967.

Left to Right:
Mulla Aqajan Bassal,
Mulla Mayer Ambalo,
Mulla Chai Cohen

Left to Right: Molla Rafael Gol, Mulla Reuben Ambalo, Mulla Moshe Naamat

CHAPTER 16

My cousin, Chai Cohen, put his arm over my shoulder and said to me, "Michael, here are the tickets. From now on, you are leading your family." He guided me to what had to be done and enhanced my confidence as a mature person who was qualified for the task.

I responded seriously, "You may trust me. With G-d's help, I will fulfill this mission successfully."

After we separated from the Jewish community in Hirat, we were accompanied by Chai Cohen until the border checkpoint between Afghanistan and Iran. On the right-hand side was a stage covered with an awning. Opposite this was a small office occupied by a border controller who was a young Afghan man. He got out of his office and instructed us to put all of our luggage on the stage and wait there. Then he asked for someone from the family to pick up all the passports and follow him into the office. I was the one who fulfilled his request.

Entering his office, I suddenly saw Chai Cohen behind me telling the controller, "We heard a lot of good things about you, especially that you liked to help people. We wanted to bring you a gift but did not know what." The man responded, "Never mind. Everything is okay. I just need to know where you are going and for how long." I said that we were going to attend a wedding in Tehran and would stay there with relatives for about a month. This man knew well that every Jew that left from this point never returned back to Afghanistan, but he did not want to cause us any difficulties.

He said to us that he had to fill out a form with details of our trip, but since he did not remember where he put his pen, he asked for mine. He noticed that I had an expensive pen in my pocket which I handed over to him. Then he asked how many of us there were. I said three adults plus five children. I gave him my passport along with my parents' passports. All of the children were mentioned on my mother's passport.

After a short while, he finished filling out all the paperwork and went with us to check the luggage since he was also in charge of customs. Then he said, "*Baman Choda* and have a nice trip." I noticed that instead of returning my pen, he put it in his pocket so I said, "I am giving you this pen as a *Yad Gari*." He said, "Oh…Thank you very much." He opened the border gate and we crossed it with our luggage to Iran. We boarded a bus to Mash'had while seeing Chai Cohen waving goodbye from the Afghan side of the border. He looked very happy since the whole process of leaving the country had passed smoothly.

The bus traveled quickly and by noon reached a customs checkpoint, which we were not expecting. A tough guy approached us and instructed all of the passengers to get off the bus with all luggages. We entered a large room where he ordered us to open all of our suitcases. He asked us what we had inside and where we were traveling to. I answered that everything was personal items and that we were going to a wedding in Tehran via Mash'had. I had not yet opened my personal suitcase and he urged me to do it. My mother got scared that this man was going to give us a hard time, so she intervened. She said, "Sir, we have a wedding in Tehran and this is my son who is the groom." I opened my suitcase without hesitation and he looked in and saw just books and albums which calmed him down. This customs controller wrote down in my father's passport that we were carrying personal items and allowed us to return to the bus to continue with our trip. The suitcases were placed on the roof of the bus again.

The trip to Mash'had was a long one, even though it appeared to be close on the map. The other passengers besides us were tourists who either read newspapers or slept. My parents and the younger children were sitting in the front of the bus while my sister Tamar and I sat in the center. I noticed an American tourist holding a copy of the weekly Time magazine in his hands and on its front cover was a picture of Moshe Dayan with the headline "Person of the year." This was only a few days after the Six-day War, and we as Jews felt full of pride and confidence. The tourist continued to read the magazine, while I told my sister that I was very curious to see what was written in it. After the tourist had finished reading, I approached him and asked him very gently if it was possible to borrow it. I did not show any signs of happiness for the defeat of the Arab countries even though this tourist could guess that we were Jewish. He smiled at me and gave me the magazine.

My sister sat beside me looking eagerly at the magazine, enjoying the many historical pictures of the battlefields. One of the photos which drew most of our attention was a huge picture of General Moshe Dayan, on the first page depicting him as a great hero who contributed tremendously to the victory of the Israeli army. Another photo showed Egyptian soldiers under aerial attack leaving their shoes to easily get away from their burning tanks and armored vehicles. All of this occurred in the Sinai desert and the Israeli flag could be seen waving on the bank of the Suez Canal. Another emotional picture showed enthusiastic Israeli soldiers near the Wailing Wall surrounding the chief military Rabbi Goren who blew on a Shofar. This was a symbol of the ultimate victory over the enemies.

A short while before reaching the city of Mash'had, the bus was stopped again for inspection by a policeman. He interrogated us and wanted to check all of our luggage. Fortunately, my father was able to show him his passport with the remark of the previous customs controller, that we were carrying just personal items. It was enough for the policeman to be convinced that everything was okay.

For the remainder of the trip, I took a nap like most of the other passengers until we reached the city at around noontime. Since our family were the majority of the bus passengers and loaded with a lot of luggage, the driver offered to take us to our final destination.

We showed him the address at a commercial center and he drove us there and helped us to unload our luggage on the sidewalk. The man who was supposed to welcome us in Mash'had was absent and his store was closed. We waited for almost an hour but there was no sign of that man. We could not believe he would be so irresponsible and we did not know what to do.

Suddenly, a nearby store owner approached us, as he saw people with luggage standing desperately for over an hour. He asked us who we were waiting for and if we needed help. We told him that we were expecting his neighbor to welcome us but the store was closed. He answered that if he was not here by now then it seemed he would not come at all and again offered his help. We told him that we had just arrived from Afghanistan on the way to Tehran and that his neighbor should have helped us in this regard. "What is the problem?" he responded in surprise. "I will show you the way to the railroad office to buy tickets. Just one of you should follow me right away since the next train is leaving to Tehran in two hours." I told my father to wait with the luggage until I came back with the tickets. I thought to myself, this man must be a messenger of G-d sent to help us in this unexpected situation.

I understood his Persian language even though it was a bit different from Afghan and I followed him closely so I wouldn't lose him. He didn't go fast and his hands were behind his back; one of them was holding *Tasbe*. After a short while of walking, we reached a beautiful mosque covered with a huge dome that was painted in blue like the sky. It was the time of prayer for the Muslims and many people were entering the mosque. At the front gate, each one of them bowed before going inside. My companion told me that we had to follow these people and pass the mosque since on its other side was the ticket booth for the train. I bought eight tickets, three for adults and five for children, and within half an hour, we returned. I hailed a chauffeur with a minivan who drove us to the central train station which was nearby. Porters helped us to load our suitcases onto the train.

A long ride of sixteen to eighteen hours was ahead of us and the very long train which we had boarded was overcrowded. In the front car that we had picked, there were not enough available seats for all of us to sit together. My parents and four siblings found seats at the front while my little brother Avraham and I were still looking for seats for ourselves. Finally, I found one vacant seat at the rear of the car. I sat there and my five-year-old brother sat on my lap. I was happy that none of us had to stand for such a long trip. The train left the station. Before boarding the train, we sent a telegram to our agent in Tehran, a very nice Jew who was well-known for his kindness and assistance to every Jew leaving Afghanistan. His name was Mr. Tadayon. We notified him of our expected arrival at eight o' clock in the morning of the coming Friday. As the train started to move, Mom, as usual, became worried if we had enough food with us for this long journey. She checked over and over again the bags of food to be sure enough was left for us, and let us know in detail: yachny, plenty of bread, as well as dried and fresh fruits.

In this part of Iran, people are not friendly and indifferent. We were very tired and pretended to also be indifferent. Since this was a very long ride, we spent the time reading books or taking naps. Fortunately, a couple behind me left the car to another one so I could put my little brother on the next seat. He slept most of the time. From time to time, I stepped forward to check on my parents and siblings. The time passed quickly and with no special difficulties, we reached the city of Tehran. The train entered a tunnel and decelerated until it came to a stop. I moved with my brother to my parents at the front of the car and saw outside the train station a man with pale skin and a white shirt. I said to my Dad, "I have the feeling that this is Mr. Tadayon." We got off the train and the man immediately approached us saying, "Are you the Cohen's?" "Yes. That is us. Very nice to meet you." I said. "Follow me," he said to us, "Today is Friday and there is not much time. My van is parked outside the station." The porters and I unloaded the suitcases from the train directly to the van.

Our companion drove quickly with us to the center of the Jewish Agency in Tehran called Beheshteyh. This was a giant building surrounded with walls and huge gates. Reaching the place, the guardsman interrogated us before opening the gate. He understood that we were new immigrants going to the land of Israel. When we entered the complex, we received an apartment with ten beds in it. The representative of the Jewish Agency who took care of us was a tall Israeli with short pants, apparently a *Kibbutznik*. He welcomed us very warmly and said that we would stay in the complex until we left to the land of Israel. Before leaving us, he told us to unpack and to come to the dining room for some refreshments. Mr. Tadayon left to his home and returned after an hour with two bowls of food for Shabbat. One was full with *palow* and the other was full with chicken soup and *gondy*. We later on met the Israeli again and he referred us to the clinic for physical exams. We asked him how long we would stay here until moving. He answered that so far it was impossible to determine since Israel was recovering from a war and our stay could take as long as three months. "You must wait with patience," he added.

On that Friday, we managed to pass all of the physical exams and got some immunization injections. We asked the Israeli man to send a telegram to Uncle Isachar, Mom's brother, to notify him that we had reached Tehran and next week we would hopefully know more details about our time of arrival to Israel. On Saturday, we did not go out but rested all day. We needed this sleep after the last two weeks of stress and wandering.

On Sunday morning, the man from the Jewish Agency came and told us that we had to go to the Israeli consulate in the center of Tehran to sign a few documents. Dad immediately became distressed and warned us not to sign any document. He said, "I think they want to collect money for our travel costs upon our arrival in Israel." I calmed him down saying, "I do not think so. No one could demand from us anything upon arrival. On the contrary, the state of Israel is spending a lot of money in order to enable each Jew to come and settle there."

We all left the complex by foot until we stopped a taxi. We showed him a note with the address of the Israeli consulate, and after a twenty minute drive we arrived. It was a very nice neighborhood with private houses in which one of them was the consulate. The consul accepted us very warmly, asked a few personal questions, and asked us to sign a document. I asked him what it was and he answered, "It is nothing special, but just a proof that you have been here for a short personal interview." We asked him when we were going to leave for Israel and he answered that it might be this week if there were no last minute changes. We left and toured the neighborhood on foot. After a short walk, we reached a commercial center and to our surprise, many of the stores were owned by Jews. We had nice conversations with them and exchanged information and opinions.

After we got tired of walking, we looked for a taxi to go back to the complex. It was very difficult to get a taxi, since when one stops many people just open the door and jump inside. After realizing that all passengers got into the cab from the other side of the driver, I tried to be too clever by going through his side to get in. The driver shouted at me, "Don't you know that the door on that side does not open?" I then returned to the sidewalk ashamed. The driver immediately understood that we were foreigners in this country and invited us into his car. We told him where we were living and he took us over there. When we reached the place, we said to him in our language, "Here we are," but he did not understand and continued to drive. We shouted and he asked in surprise, "What happened?" We explained to him that he had passed the place and then he realized why we were upset. He told us that it was not his fault since in the Persian language you have to explicitly say *Negahdar! Stop!* He then made a u-turn and returned to the gate of the Jewish Agency's complex. We got out of the cab and entered our place of residence.

The next morning we returned to the consulate by foot. We wanted to check if there was any news regarding our travel. The officer there said to be patient, as the day of departure was getting closer. On Tuesday, my mother did our laundry by hand with soap she brought from Afghanistan while we were resting in our room. Suddenly, the man from the Jewish Agency came in and notified us to be prepared for four am tomorrow morning to leave for the airport.

It is difficult to describe the tremendous excitement and happiness that befell on us, especially for Mom, who said at that moment that the minute we arrive in the holy land she would kiss the sand. She prepared all of our holiday clothing and after we packed all of our suitcases we went to sleep very early.

It was difficult for me to fall asleep because my head was full with thoughts. What would it be like in the land of Israel? Is it really as people described it? A wonderful, beautiful country, with strong people filled with self-confidence? Do they believe like me that the recent victory was a miracle from G-d, or do they see it as natural that a tiny state like Israel could defeat millions of enemies within six days of war? For the last few days, we got all the news only from the Iranian radio which quoted the false news from the Arab countries. I was very eager to be in Israel already and to hear the real stories from friends and relatives who experienced those historical events. I knew that within a few hours we would be there and I did not have patience for the time to pass by.

I think that I didn't sleep at all that night, since in the middle of my thoughts I heard the steps of the man from the Jewish Agency getting closer while he shouted "*Boland Shin*". I looked at the clock and it was almost four am. As usual, Dad was the first one to get up from bed and to urge us to do the same. Anyhow, within ten minutes all of us were ready to leave. A tiny truck entered the yard, stopping at our door. Our luggage was loaded onto this truck and then our whole family got on.

My parents and little siblings sat in the cabin with the driver while Tamar and I were behind with the luggage. When we left the complex it was still dark and a cool and pleasant air embraced us. We passed the main streets of Tehran and got on the highway leading to the airport. The little truck was overloaded with our big family and luggage, and in the middle of our journey, one of its tires exploded. The driver pulled over and did his utmost to change the tire as soon as possible. We could not afford to be late at the airport. Anyhow, it took him only fifteen minutes to accomplish his task and we sped towards the airport. We soon saw the little blue lights of the airport runway since it was still dark outside. The truck entered the airport and stopped at the front of a building where a few people stood outside. They were officials of the airport who instructed us to get immediately into the building where at the back was an El Al airliner waiting for us.

Within a few minutes, we found ourselves on the airplane. To our surprise, it was totally void of passengers and we saw only two stewardesses who told us that we may sit wherever we wanted. I took a seat near the window since I wanted to see the view outside. The sun started to rise and I noticed that a missile was attached to the right wing of the plane where I was sitting. It seemed to me that a precaution was taken by the Israeli airliner management to protect the plane during this period of tension. Right before takeoff, we heard the captain announce, "Welcome aboard El Al. We are going to leave now for a nonstop flight to Tel Aviv. Please feel at home." The plane took off and right after, the stewardess served us with cold drinks and light refreshments. This was the first flight in all our lives so it is understandable that we were very excited.

Dad recited Tefillat *Haderech* and we sat quietly during most of the flight. The pilot and crew were also silent and did not report anything during the whole flight. When the plane started to descend we were already over the Mediterranean. The view was fascinating when we suddenly saw the buildings of Tel Aviv from a low altitude, and within a few minutes the landing gear of the plane touched the ground and it came to a full stop.

The day was Wednesday the twenty-eighth of June, 1967. As we got off the plane, Mom immediately kept her promise and bent down to kiss the ground of the holy land. We entered a terminal where two officials welcomed us. One was from the ministry for absorbing new immigrants and the other one was from the state housing company called *Amidar*. The first officer asked us for our names and dates of birth, but we knew only the Hebrew dates. He also did not know the corresponding regular dates so he guessed them. He then issued each of us a personal certificate of a new immigrant. My father came out ten years younger and my six-year-old brother Avraham became four years old on his certificate. All this was done quickly, and in the view of that officer, the exact age of a new immigrant did not play an important role. For him the main point was that we were back in our homeland and he wanted us to think the same.

It was interesting to try to converse with these people without understanding each others' language. They did not understand Afghan and we did not understand Hebrew. The officer of *Amidar* was an Iraqi born Israeli who told us to follow him and he guided us to a truck which was waiting outside of the terminal. He helped us load the luggage onto it and urged us to get on. I asked him where we were going and he answered, "An apartment is waiting for you in a small town called Yahud, known to the Muslims as Yahud." I instantly responded, "What? We want to live in Tel Aviv near our grandparents and uncles." He said, "Do not worry. It is just a ten minutes drive from Tel Aviv." I said, "My uncles are expecting to come here. Let's wait a bit for them and then we will decide together." The officer answered, "No way. You will wait for them in the apartment."

He immediately instructed the driver to leave the airport to Yahud which was a ten minute drive. This man was tough and did not leave us any choice but to listen to his instructions. When the truck entered the suburb, it stopped near a small office building and the officer got off saying, "My office is here and it will take me a few minutes to bring the keys for your apartment." He returned and within a few minutes we reached our designated building. The first impression we got from this suburb was not encouraging at all. All the roads crossing it were not paved but just sand routes. Most of the land was covered with orange orchards with a few factories near them. We could see a few buildings especially built for new immigrants, and here and there were several private wooden houses. At least the highway between the airport and the city of Tel Aviv passed near this suburb so we could have easy access to our relatives.

The apartment we received was on the First Floor in a new building that already housed another two families. They were also new immigrants who had arrived here right before the Six Day

War. They were privileged to experience those historical days in the land of Israel which we had missed. The neighbor opposite our apartment, Mr. Schwartz, heard noises of luggage being moved so he opened his door and asked us where we came from. He spoke only Russian and fortunately, my Dad could converse with him a bit in this language. Dad told him that we had just arrived from Afghanistan and he instinctively responded: "Afghans? These are very good people. I met a few of them in Russia." He and his wife were a cultured and quiet couple who were the first people whom we befriended in this country. We entered the apartment with all of the luggage and checked it thoroughly.

The apartment had a small living room and another two bedrooms, a kitchen, and a porch overlooking an empty field. A bit further, we could see an industrial zone surrounded with tall eucalyptus trees. This was a very small apartment for such a big family like ours, especially since my brother Amnon had decided to live with us after being separated for fifteen years. At least the officer of *Amidar* made sure to supply the apartment in advance with beds and blankets.

It was summer time and by noon of the day of our arrival, we finished all of our arrangements in the apartment. Now we had to figure out how to get in touch with our uncles, and my siblings Amnon and Esther. Naturally, this would have been a very emotional reunion after so many years of separation. We started to feel hungry and Mom sent me out to buy some food and drinks.

Fortunately, I met a young guy of Iranian origin with whom I could communicate in Persian. He accompanied me to a nearby grocery and suggested to buy a few sandwiches and drinks. Interestingly, I didn't know for what the Israeli's called a 'sandwich' until the grocer prepared a few for me according to my request. He cut a bun into two halves, placed between them sausages with humus, and put the halves together. I came back home with the food and all of us ate with a great appetite. After satisfying our hunger, Dad looked at me and asked, "What is next? How are we going to meet our relatives?" As usual, I calmed him down and offered to go together with my brother Ishai to look for our uncle Isachar's address. Dad said, "Don't be silly. Tel Aviv is a big city and you are not familiar with it at all. You will just get lost." I answered, "Don't worry. Leave it to me. We will find him."

We left home and showed people in the neighborhood the address we were looking for. It was in a district of Tel Aviv called Kiryat Shalom. No one there knew the address but we were advised to take the bus to Tel Aviv and to ask people when we get there. While waiting at the bus stop someone working in our neighborhood passed by with his car and offered us a ride towards Tel Aviv. On our way, we conversed with him a bit in English and showed him a note with the address we were looking for. He did not know where it was, but at least he was bringing us closer to Tel Aviv. He drove to a gas station and dropped us off. Later on, we learned that our sister Esther, who immigrated to Israel in 1953, lived in Salama Gimel, which was a few minutes walk from this gas station and we didn't know. We asked the employee at the gas station, "Do you speak Persian?" He answered no. At that moment, a cab driver pulled into the station and we asked him the same thing. He also did not know Persian but after looking at the note with the address on it he said, "Get in. I know where it is and I will take you there." I had ten *lira* which I had received from the man at the Jewish Agency, so Ishai and I felt comfortable getting into the cab.

After a ten minute drive, we got out of the cab, right in front of the house where our uncle Isachar resided. We knocked on the door and the children of our uncle opened it. This was during the period of the school summer vacation and all of them were at home and looking at us in amazement. We were dressed in fancy clothes as if for a holiday and did not have a common language with our cousins Mordechai, Blorya, and Osnat, whom we had just met for the first time. We stood and stared at each other without saying anything. Suddenly, their mother appeared from another room and said, "Welcome. Please follow me into the living room and take a seat. My husband Isachar and your other uncle Nafthali had just left a short while ago to look for you at the airport. If you want, I can take you to grandpa and grandma who live across the road." "Okay," we answered and all of us left the house.

Grandpa was already standing outside his house impatiently expecting our arrival. We hugged and kissed each other. He said, "Grandma is inside the house waiting for you. Please come in." Again, there were kisses and hugs and all of us were very excited. I've heard many stories about Grandpa and what impressed me the most was his righteousness and here I saw him for the first time in my life. I was so stunned that I could not talk for a few minutes. They asked us, "Where is the rest of your family?" We told them in brief what went on with us since coming to Israel and in the middle of our talk the door opened and it was unbelievable who we saw - Our two uncles together with Dad and Mom.

The exciting meeting among these close members of the family was accompanied with many tears especially from Mom when she saw her parents. I asked them, "Where did you meet and how did you get here?" Isachar said that he took his brother Nafthali on his three-wheeled motorcycle to the airport where they got Mom and Dad's new address. They traveled right away to Yahud, met my parents, and then all of them came directly here.

The rumor of our arrival spread among the members of our family and in a short while my brother Amnon arrived accompanied by my sister and her husband Levi Simantov whom I met for the first time. Old friends of my parents who immigrated to Israel many years ago also heard of our arrival and came to see us. I had no idea that we had so many family members and friends here. All of them spoke our language and I felt like I was right at home in Kabul. They welcomed us very warmly and were eager to know how life in Afghanistan was recently. All of them left the country in different periods of time since 1948, the year the state of Israel was established. The absolute majority of them came to reside in Kiryat Shalom which was a southern district of Tel Aviv and they tried to convince Dad to join them. Dad responded that it was too early to make such a decision. He said, "We have just arrived in this country and we have to look around and think it over."

We were lucky that my siblings Esther and Amnon grew up in

Israel. They spoke the Hebrew language fluently and were familiar with the mentality of the Israelis which was unusual to us. We got help in taking care of our interests with the Israeli institutions from them. My sister Esther, despite being eight months pregnant, went to different offices on our behalf for our smooth settlement in the new country. For example, she went to an office of the Health Ministry to get health insurance for us. She also registered the adults in the family in an employment agency and our little siblings were registered to schools.

My brother Amnon was more helpful to us with his knowledge of all aspects of Israeli life which he shared with us. He was very busy at his job as a soccer coach for the Israeli army. He had to work seven days a week so he barely had free time for a private life. The team of he trained was called Gadna Yehuda.

CHAPTER 17

During our stay with our grandparents, my uncles made nonstop attempts at convincing Mom and Dad to move near them, but my siblings Amnon and Esther opposed this idea. They loved the neighborhood of Yahud, which was a small developing town, so more work was available there. Dad personally liked Yahud because it was outside of the city and therefore a quiet and tranquil place with clear air. Grandpa asked us to stay with him for our first Shabbat in Israel and we immediately agreed. There was no problem housing us since many of our relatives lived in the area and our family split up to different houses.

I went to stay in a vacant room that uncle Nafthali offered me at his house. For the Shabbat prayers, we went to a nearby synagogue filled with Jews from Afghanistan. All of them knew Dad and were very happy to see us there. The prayers in the synagogue were accompanied with melodies that were a mixture of styles from Afghanistan and Jerusalem.

After the Shabbat meal, my brother Amnon suggested that we take a tour of the city. The weather was nice and warm and we went slowly by foot to see the central bus station of Tel Aviv. It took us just ten minutes to reach it. The station was crowded with people, especially soldiers who could be seen everywhere walking very proudly after their victory. Some of them were carrying submachine guns under their arms, and other soldiers who were on vacation were walking with their girlfriends.

From time to time, Israeli combat planes were seen flying over the city as well as military vehicles passing down the streets. The big war was already over but there were still lingering clashes near the borders, especially with Egypt. When Mom saw some people smoking cigarettes at the central bus station, she asked my brother if they were gentiles. Amnon responded, "Why do you think so?" Mom answered, "You see that they are smoking on Shabbat." Then Amnon explained to her that they were Jews, but in Israel many people observed the religious laws to a lesser degree than elsewhere. My mother could not accept this.

We left the station and reached the main commercial street of Tel Aviv, called Alemby, named after a British general. We walked down the street towards the beach. Mom was very happy to see that most of the stores on both sides of the street were closed to honor the Shabbat. We then saw the Mediterranean in front of us which was the first time I saw the ocean up close. This tour took us more than two hours until we returned home to rest for Shabbat.

We stayed in Tel Aviv with our relatives for several weeks. Each evening we were guests at another relative's home. It felt wonderful being with them and they did not want us to leave, as long as the apartment we had just received was still void of furniture. My parents used their stay in Tel Aviv to buy furniture, electrical appliances, and more for our new apartment.

Two weeks after our arrival in Israel, we received an invitation for the wedding of a cousin. This was an occasion for us to meet our whole family including the members we had not yet met. I was especially eager to see Uncle Matityahu, the groom's grandfather, about whom I had heard many stories from my parents. He was the head of the Jewish community in Hirat, and met the king Aman-Ula Chan in the year 1920. I also wanted to meet Grandma Miriam, the nanny who raised my father.

This was the first wedding that we attended outside of Afghanistan and we were very impressed. It took place in a luxurious hall with a catering service, waiters, and an orchestra with electrical instruments. The music was of Greek and Israeli origin which we were not accustomed to but enjoyed very much.

Suddenly, Mom approached me and said, "This man is Uncle Matityahu." He was tall, had an impressive appearance, and looked strong and healthy despite being eighty years old. While I was looking at him in astonishment, my aunts laughed at my response. I asked them why they laughed, and they said, "Do not look at him as an old man. During all his eighty years, he visited a doctor only once just a few weeks ago and never took any medication. The medical staff at the hospital could not believe such a phenomenon. Would you like to talk to him?" I agreed immediately. Aunt Shoshanna introduced me to him and said, "Michael is the son of my sister and he recently arrived from Afghanistan and wants to say hello to you." He said, "Hello my son and welcome." Then he turned around and left. I understood his attitude, since he was a tough man, and he addressed me as a child.

Next I met my uncle Sasson for the first time. He was Dad's stepbrother, and the son of Nanny Miriam. Sasson was very happy to meet me and enjoyed talking to me. He told me the story of his family's arrival to Israel in 1950. He was ten years old and his father was deceased. The state of Israel had just been established and did not have enough resources to absorb the new immigrants conveniently. "We had been referred together with five or six other families from Hirat to a neighborhood near the city of Netanya where all the inhabitants lived in tents. The place was called a *Maabara*. We arrived there on a Friday a few hours before Shabbat with no food or money and were thrown straight into a tent. An official from the Jewish Agency brought us just a bit of bread and margarine and that is how we welcomed our first Shabbat in Israel."

Uncle Sasson continued to tell me in detail how he and his family stayed and struggled to survive in the Maabara for three years. His mother went every day on foot to a nearby small town called Ra'anana to work as a maid. He saw how she worked so hard and brought home very little money, so he decided to help with making a living.

One early morning, he went on foot to Ra'anana and met a man who rode on a horse and wagon. The man asked him, "Child, are you looking for work?" and he answered "Yes." Then the man said, "Get on the wagon and join me." "No," my uncle answered. "First I have to go home and tell my mother that I am starting work so she will not be worried, and besides I also have to bring food from home." The man insisted that he needed him right away for work and there was no time to go home and tell his mother. Sasson had no choice so he got on the wagon and they rode for about fifteen minutes on narrow paths until they reached the man's wooden house. The house was surrounded with orange trees and vegetable gardens. He asked his wife to give Sasson something to eat so they could start their work day. He instructed my uncle to follow him and his horse which was now connected to a plow and to gather all the potatoes into a bucket. When the bucket was full, Sasson took it to a huge box to pour the potatoes in. He performed this routine over and over until they finished the whole field of potatoes by evening. Then the man told my uncle to separate the rotten potatoes from the good ones and when he finished this task, he paid him a good daily salary of two liras.

He praised him for his good work and asked him to come tomorrow in the early morning again. Sasson was taken by his new boss to the main road and from there, he happily ran home. His mother asked him, "Where have you been the whole day? I was so worried about you." He responded proudly, "This was my first day of work and here are the two liras I earned." Sasson's mother was upset, she did not like the idea and said to him, "You do not have to work now; you are still a little boy." The next day, he woke up early and went again to his work. This lasted for fifteen days and at the end, he gave the whole amount of thirty lira to his mother which helped her a lot. She was very grateful.

From an early age, my uncle liked to work. When he was thirteen, he joined a group of older friends from the *Maabara* who traveled daily to Tel Aviv for work. When he reached an industrial zone in southern Tel Aviv, he saw a sign on an ironworks factory saying that workers were wanted. He got in and was immediately denied by the owner who said, "You are too young to work here. You are still a child." My uncle insisted that he was capable to do any work like an adult and he told the owner that his father had died, and at home were his mother and two sisters. He had to work so the family could make a living. The owner looked at him and said, "Where do you live?" Sasson answered that he was living in a *Maabara* near the town of Ra'anana. The owner responded, "Oh, Ra'anana! No way can I hire you. We start at seven o' clock in the morning and you could not make it so early." My uncle said with full confidence, "Do not worry, I will be on time. Just try me." When the owner saw that the boy was very serious, he decided to give him a chance and hired him just for a trial in the beginning.

The next morning, he woke up at four a.m., prayed, and caught the first bus to Tel Aviv. He reached the central bus station and from there he ran to the factory. At 6:30 in the morning, he was near the gate which was still closed. The next person to arrive was the owner who could not believe what he saw. "Did you sleep nearby during the night?" he asked my uncle. "No, I came from home with the first bus and I have been here since 6:30 in the morning."

My uncle then turned to me and said, "Since that first day of work in that plant, I did not miss even a single day excluding Shabbat and holidays during my fifteen years of work there. Over time, I learned the job very professionally and the owner, who realized that I was diligent, appointed me as the manager of the plant. I earn now a very good salary and if your family needs help, I would be ready to assist. Contrary to my family, you arrived in Israel in a very good time of prosperity so I am sure that your absorption in the new country will be much easier than ours."

I thanked my uncle for his nice offer of assistance and we separated. On that night of the wedding, we met almost all of my family members in Israel. They had left Afghanistan in the 1950's and approached us in curiosity with many questions regarding the country they left. We told them that we were happy to arrive here in Israel despite leaving Afghanistan where we had a very nice living. We did not leave because of persecution, but just wanted to come and settle in our Jewish state. We have good memories from Kabul where we lived for many years and feel grateful for the wonderful people, Jews and non-Jews, of that city.

Some of the family members were my age and we did not have a language in common. They did not know the Afghan or English languages, and my Hebrew was not fluent enough to converse. On that evening, I finally decided that I must be thoroughly familiar with the Hebrew language as soon as possible if I wanted to succeed in this new country.

We lived with our family for almost three weeks. Dad had already become impatient and I could understand him. A family of ten people must live in their home, but Mom wanted to continue this way of living with relatives. They were her closest family members; her parents, brothers and sisters, whom she had not seen for so many years. Our apartment in Yahud was already furnished along with electrical appliances and utensils.

I told my parents, "Let's celebrate our first Shabbat in our own home this week and invite Grandpa Yehuda and Grandma Sara." I almost succeeded to convince my mother with this suggestion, but my uncles insisted that we must stay with them until after the upcoming high holidays of *Rosh Hashana*, *Yom Kippur* and *Sukkos*.

We were already in July in the middle of the hot summer of Tel Aviv when Uncle Isachar brought two tickets to an organized tour in the old city of Jerusalem. He got these tickets from his place of work and asked me to join him. I was very excited since it was an opportunity for me to see all the holiest places of our religion which had just recently been captured and returned to Jewish possession. On the first scheduled day of the tour, Isachar could not come so his daughter, Blorya took his place.

We traveled on a touring bus that was full of Isachar's coworkers. The tour guide was an educated man who was fluent in English but described each site we passed in Hebrew. We crossed the new city of Jerusalem which was modern until we reached an entrance to the old city. This ancient part of Jerusalem dated back to the time of the Bible when King David ruled the land of Israel. It is surrounded by a tall wall which has several gates to enter the old city.

The bus parked in a lot near the Jaffa gate. There I saw many people getting off their touring buses and flocking through the gate into the old city. Security police of the Israeli army were in control of order in this part of the city. Before passing through the gate, our tour guide warned the members of the group to stay close together otherwise, we would have gotten lost in this tremendous crowd.

My cousin Blorya and I passed together through the market which was composed of many stores with a wide variety of merchandise. Arabs were also wandering among the crowd in the narrow alleys and it was impossible not to lose the group. All the stores belonged to Arabs who welcomed the tourists very warmly and made a lot of money from them. We went very slowly because of the crowd until we reached about a hundred yards away from the Wailing Wall. It was impossible to get closer to the wall since thousands of people were there. Near us, another tour guide was standing and elaborating in English to his group the history of this place.

Blorya and I were stuck there and listened intently to his speech. After a few minutes, we looked around and could not find our group. I wondered, "Where did they disappear to? I had just seen them and now they are gone." We looked all around us for our guide and Blorya asked people in Hebrew but nobody could help. We decided to make a U-turn and go back the way we came from. We could hardly advance through the market while looking for our lost group. We searched for about two hours until we reached the Jaffa gate. We knew that the bus was parked nearby and we needed to get back to it. The whole group was already seated on the bus and the guide welcomed us with screams. "Where did you disappear to? All of us are waiting for you and now you remembered to return?" I answered, "I apologize sir, we simply lost the group." As we took our seats the bus driver started the engine and returned directly to Tel Aviv.

My whole family waited impatiently for my return since they had decided on that evening to move to our new home in Yahud. My brother-in-law Levi told me that Esther was already there since the morning. She went to the local office of the Ministry of Education to register my little brothers. Rafael and Gavriel were registered to elementary school, while the smallest, Avraham, was registered to kindergarten.

He told me that the family was considering referring me to study in an *Ulpan* on a *kibbutz* far from home. He explained to me that this was a Hebrew course which takes a few months and it was the best way to learn the language quickly. After I mastered the language, I could choose a vocational school where I would learn a profession and could make a nice living. I listened to him carefully and then said, "This plan sounds very good, but what will happen to my parents? How could they make a living when I would be away from home?" Levi said, "Don't worry for them. Your older brother Amnon has a job and will assist them financially. Besides, your father is not so old yet. He is only fifty-five and can go out to work." I said to him, "Okay, you have convinced me. But now the question is what my parents would say to this plan." I told my parents about my discussion with Levi and asked for their opinion. Their instant answer was, "Do what you think is good for you." We then drove to our new home in Yahud.

The day after, I was accompanied by my sister Esther to an office of the Ministry of Immigrant Absorption. It was a nice favor on her part since she was already in the ninth month of pregnancy and it was hard for her to walk. I was registered to attend an *Ulpan* on the *Kibbutz* Mesilot which was far from home. This *kibbutz* was located between the towns of Afula and Bet She'an, not far from the Galileo Lake. The officer instructed me how to get there and told me that the class would start in one week. I returned home.

Now I had time to check our new apartment in detail. I realized that no essentials were missing and my parents and siblings could enjoy living in this apartment. Dad asked me to accompany him to the local market which was a ten minute walk. It was a small market and the only commercial center in this little town of Yahud. There we bought our food, vegetables, and meat. The majority of the people there were Jews from Turkey. Dad tried to talk with them in the Uzbek language which is similar to Turkish but they did not understand him. We passed a coffee house and saw there the Jewish Turks sitting and playing *sheshbesh* while drinking arak instead of coffee.

In Yahud, there were almost no motor-vehicles and no paved roads. Only the main street which crossed this little town was paved. The other roads were just plain sand. There was only one branch of Bank Leumi which was also located in the market. Dad and I made all of our purchases, carried the food in baskets, and returned home. Mom started her usual cooking and said, "That's it. The party is over. I am back on track to work in our home."

In the last week left for my stay at my parents' home in Yahud, I decided to go to Tel Aviv to meet some of my childhood friends from Kabul, Amnon and Elimelech. They wanted to take me to either the beach or a public swimming pool. We went to the beach in Tel Aviv and after a short while, decided to go to a nearby swimming pool called Gordon. We bought tickets to get in and then each of us got a locker for clothes. For the first time, I saw people getting undressed and walking totally naked. It looked very weird to me. Amnon, who noticed that I was embarrassed, said to me, "Don't be shy. All the people here are men. Undress and put on your swimming trunks like all the others. In Rome, do as the Romans."

This way of behavior is unheard of in Afghanistan. The Jews there as well as the non-Jews were modest, shy, and bashful in such situations. I encountered the same lack of modesty while sitting on public buses when suddenly a young girl would get on the bus and sit near me. All of this embarrassed me because I was not used to it, but it didn't take much time to realize that I was in another country with different customs.

The coming Shabbat was the first one to be kept at our own home in Israel. That Sunday, I had to leave for the *kibbutz* for three months of extensive Hebrew studies and then I would decide what to do next. On Shabbat, we went to the central synagogue of Yahud, which was a five minute walk from home. All of the Jews there were immigrants from Turkey but to our surprise, their prayer had the same style as the Jews in Kabul. My brother Amnon introduced us to the spiritual leader of the congregation, Rabbi Elazar Ha'Levi and to the cantor, Mr. Gavriel Bar'Simantov.

They did not know much about the Jews of Afghanistan but

welcomed us very warmly. In our neighborhood, our family was the only one from Afghanistan and there were just a few families who were immigrants from Tunisia, Libya, and India. There was no common language between us and the other residents but my parents got accustomed very quickly to their new place and neighbors.

On Sunday morning, I left home with a backpack and five liras in my pocket. My brother in law Levi accompanied me to the central bus station of Tel Aviv and we looked together for the bus leaving to the city of Afula. It was a two-and-a-half hour drive. From that city, I was supposed to take another bus to reach the *kibbutz*. On that day, the central bus station was overcrowded with soldiers who had returned to their military bases after spending Shabbat with their families. We were already six weeks after the Six Day War, and still military vehicles and soldiers with submachine guns were seen everywhere. I looked at them and said to myself, "In one year, I will join the Israeli army too and I hope that by then there will be real peace between Israel and the neighboring Arab countries so I will not need to fight." While I was thinking of this, Levi found the bus stop to Afula and bought for me one ticket. He also asked the driver that when he would arrive at Afula to guide me to the bus stop from which the bus to *kibbutz* Mesilot left. The driver calmed down my brother-in-law and said to him, "Don't worry, I'll take care of this young man." I embraced Levi and we separated.

I got on the bus, and after three hours, I got off the second bus near the gate of the *kibbutz*. The guard interrogated me for a bit and I showed him the official referral letter from the Ministry of Immigrant Absorption. He showed where the office of the *kibbutz* was located and he let me in. A nice woman welcomed me there and filled out a form for me while asking me private questions.

The *kibbutz* was comprised of small houses and she assigned to me a room in one of them. Then she instructed me to go eat lunch in the communal dining hall of the *kibbutz*. When I arrived there, I met my future Hebrew teacher. She said to me in Hebrew, "Hello. What is your name?" I did not understand her and she asked in a different way, "How do people call you?" "Ah...Michael," I answered. Then a sequence of short questions and answers followed: "Where are you from?" "Afghanistan," "Are there Jews in Afghanistan?" "Yes, like me." "Do you speak English?" "Yes, little." "Where do you live in Israel?" "Yahud." "Where is Yahud?" "Not far from Israel's international airport." "Okay," she finally said. "It is two o' clock in the afternoon and I guess that you are very hungry." I answered, "Yes." For the first time, I saw people standing in the line to put food on their trays and then moving to their tables. I did the same and started to eat with a great appetite.

After I finished my meal, the teacher approached me and said, "Michael, there is a Hebrew class at four o' clock. Please come on time. You have almost two hours to rest until then." I said, "Okay" and went directly to my room. I organized my belongings in the closet and I took a nap in bed that lasted to almost four p.m. When I got up, I went outside to look for my class. On my way, I saw many small houses hidden behind tall trees and between them were lawns.

Finally, I found the class room and the teacher was already lecturing. When she saw me, she said, "What happened to you? Why are you late? Please come in." I entered the classroom and took a seat. There were another fifteen students, males and females, from various countries like Poland, Argentina, America, India, Iran, and more. My knowledge of Hebrew was poor, besides it being a very difficult language to learn. The letters are totally different from any other language and so is the grammar. The teacher spoke very clearly but there were many words she used which I did not understand. I had in two dictionaries in my possession, one Hebrew/English and the other English/Persian, which I used many times during the lesson. I listened very carefully to everything she said and tried my best to be a good student. I wanted to fulfill the promise I made to my sister and brother-in-law to master the Hebrew language very quickly and to return home shortly. I put a lot of effort into this endeavor.

Our daily routine was to wake up very early in the morning and at four a.m. we started to work. The studies in the *Ulpan* and stay in the *kibbutz* were not free of charge; we had to work for it. The supervisors woke us up and we gathered in the dining hall just to drink coffee. A tractor took us to a field of olive trees. Our task was to pick the olives from the trees until eight in the morning and then return to the dining hall for breakfast. Some of the students worked in the cowshed and others in the kitchen of the *kibbutz*.

It was our choice to choose which work to do and I preferred the olive picking much more than anything else. I liked this work and did it enthusiastically. One day, I wanted to pick as many olives as possible in a short amount of time. I forgot to get on the ladder and climbed on a branch to reach higher even though it was strictly forbidden. Suddenly, the branch broke and I fell from a significant height. To my good fortune, I fell on a heap of straw and did not get hurt. My friends ran to me to check if everything was okay and warned me not to climb like that again.

One day on the *kibbutz* I met a few Jews from Iran who worked in construction. They told me about a Persian girl who was smuggled from Iran to Israel and felt very lonely in the *kibbutz*. They asked me to meet her and to provide her with company. I went to her room and met a very sad girl. She was almost sick from loneliness. Unfortunately, we did not have a common language and a day after our meeting, she left the *kibbutz*.

After my early morning work at the olive orchard and breakfast in the dining hall, I studied in the *Ulpan* until 12:30. Then we ate lunch and had a break. This was during summer time which is very hot in this part of Israel. Nevertheless, we did not have air conditioning in the rooms or in the classes. After a nice break, we returned for studies from 4:30 to 6:30 p.m. and then we ate dinner. This was more or less our daily routine in the *kibbutz*.

On Shabbat, we did not work or study and had the choice to either travel home or stay in the *kibbutz*. I chose to stay in order to use all my free time to improve my Hebrew. Usually I sat on the lawn with my two dictionaries and studied the language.

I checked if their was a synagogue on the premises but I was disappointed to hear that even though there was one, it was closed since no one in the *kibbutz* was religious.

On the first Shabbat, the supervisor of the kitchen, a young red-haired man, approached me and said, "Today, is your turn to work in the kitchen." I responded, "It is eight in the morning, and I should have been in a synagogue and you are telling me to work in the kitchen now? What is this?" He responded that here in Israel, the Holy Land, we do not have to pray. He said that I did not even have to put a *Kippa* on my head since these customs were appropriate for the Diaspora to keep our Jewish identity. I could not argue with him because it was difficult for me to express myself in Hebrew as well as his opinion was unchangeable. He did not leave me any choice other than to follow his instructions. After the members of the *kibbutz* finished their meal, all the dishes were cleaned by a few of my friends in the kitchen and then I had to put the dishes back in their places. It took me hours to complete my job and the red haired man came into the kitchen in the afternoon to check on us. He could not believe what he saw, and said to me, "Michael, congratulations. We have never had such cleanliness and order in the kitchen until now."

On that Saturday evening, a Hebrew movie was planned to be screened in the dining room. I was eager to see it but before it started, my teacher came to me to ask for a favor. She said, "One of my students from India, named Yaakov, has gotten a terrible toothache and needs to urgently see a dentist. The nearest dentist available now is in the town of Bet She'an. Could you please accompany him? It is about a half hour walk on foot." I went to his room and found there an adult, in his thirties, crying like a baby. I introduced myself and we left for the dental clinic. The road was dark and he was holding his face the entire way from pain. He hardly said a word.

The dentist took X-rays of him while I sat in front of the room. Afterwards, I heard the dentist say that the tooth had to come out. Out of fear, Yaakov cried more than ever before. The dentist was unmoved by his cries and gave him an injection to sedate him. Yaakov immediately got quiet and within a few minutes, the tooth was taken out. We thanked the dentist and left.

Many times in Israel on Saturday nights, there are free performances for the public. Yaakov, who had calmed down from his ordeal, suggested to tour the town and see what was going on.

In one theatre, we saw Israeli singers and in another theatre, we listened to a concert in the park. After spending over an hour in the town, we went to the road that led to the *kibbutz*. In the middle of our way, Yaakov started to cry and scream. "What is it now?" I asked. He answered that the dentist mistakenly took out a wrong tooth. At that point, I did not know whether to laugh or to cry. "What do you want to do now?" I asked him. He insisted that we had to return to the dentist. He spoke in fluent English and I understood him very well. We returned to the clinic and the dentist calmed him down by saying that it was only his imagination. He said, "The right tooth was taken out and as the anesthetic wore off you felt pain all over that region." The dentist gave him a few tablets to treat the pain and assured him that everything would be okay by tomorrow.

We left and again spent another half hour listening to music and watching dances. We then returned to the *kibbutz* and I fell asleep in exhaustion. The next week was similar to the first. In the morning, I picked the olives. Twice a day, I had Hebrew classes and in the free time between classes, I would sit on the lawn and study Hebrew.

One of those days, the red haired kitchen supervisor looked for me. He found me sitting on the lawn as usual and as he approached I immediately stood up. He asked me, "Why are you standing straight? I am not a military commander and you do not have to do that. I just wanted to thank you for the great job you did in the kitchen." He then said a few more things which I did not understand since he talked in quick and difficult Hebrew. I did not answer but only said, "Okay" and he left me. My teacher told me afterwards that he was very satisfied with my work in the kitchen and wanted to employ me there for a steady job.

On Friday of my second week in the *kibbutz*, I missed my family terribly and decided to return to Tel Aviv for the weekend. I asked permission from the secretariat and was approved. On the bus to Tel Aviv, I decided to first visit my sister and her husband, Esther and Levi. It was almost noontime when I got to their house and was very surprised to see my parents there as well as many of our family members. The atmosphere was of joy and happiness and I immediately asked what was going on. The answer was that my sister Esther had just given birth to a son and the family is preparing for a party on Shabbat to honor the father of the newborn.

I was wondering to myself why they did not let me know, but there was no time to ask since everyone was very happy and wishing to each other *Mazal Tov*. My mother, as usual in such circumstances, was very busy in the kitchen preparing meals for the party tomorrow. Dad was overwhelmingly happy since it was his first family celebration in Israel – the birth of his grandchild. Dad was also given the honor of being the *Sandak* on the coming *Brit Millah.*

Levi welcomed me with a broad smile and said, "I am happy that you returned. We received from you your first letter in Hebrew which you sent from the *kibbutz*. I am really proud of you. Your mastery of Hebrew has advanced tremendously and I even notice it when you talk. I do not think that you have to go back to the *kibbutz*. "No, no." I responded. "I still need a few more weeks to be there in order to feel confident with the new language." Then my brother-in-law decided to go into more detail in explaining why I should not return to the *kibbutz*. He said, "Your parents are not pleased with your stay there. They have a lot of expenses and your father does not earn enough to make ends meet. He succeeded to get a job but you cannot imagine how difficult it is. It is called here *avodat dachak*. He works in Yahud digging holes where pipelines have leaked and a technician comes and fixes them.

Your sister Tamar got a job as a laborer in a factory for *halva* and your younger siblings are at home waiting for the new school year." Levi, with the help of my father, were preparing for the upcoming *Brit Millah* celebration that was scheduled for the middle of next week. My sister had already returned from the hospital in a very good mood. I stayed until the Brit of the new grandchild who was named Gili. Afterwards, I returned to the *kibbutz*. The teacher's first question to me was, "Michael, what happened? Why did you miss a few days of class?" I did not know how to express myself so I said, "Sister...baby boy...stomach." The teacher nodded and said, "Oh I see. She gave birth to a boy. *Mazal Tov*. We missed you, Michael and we are glad that you are here. Now back to our studies." I responded, "Yes teacher. I actually returned only for my Hebrew studies." While saying this, I thought about the words of my parents, "Michael, you must be at home. We need you." I remembered the disappointed feelings of my mother because of Dad's physically difficult work of digging holes in the ground in such an advanced age. It was difficult for us to accept the extreme change of my father's status. All of his life until now, he was a community leader, organizer, and guide for people. Now in Israel, he was forced by circumstances to be busy with a job that we could not even imagine.

After class, I went to sit on the lawn and thought to myself, "What advantage do I have in being on this *kibbutz*? On one hand, I improve my Hebrew but on the other hand, people here use me for extra work that I do not have to do. They just take advantage of my good heart to exploit me. I could at the same time attend a vocational school where I could kill two birds with one stone. In such a school, I can learn a trade and simultaneously improve my Hebrew, since all studies would be in the language."

On the spot, I decided to stay in the *kibbutz* for a little long, and then return home to help my family. I acted on my decision and stayed in the *kibbutz* for another two weeks and then I informed the teacher that I would not be coming back. My decision came to her as a big surprise. She asked, "Why do you want to leave? It seemed to me that you were very happy here and your progress in mastering the language is very good." I answered that it had nothing to do with the *kibbutz*, but the situation at my home. My family simply needed my help.

On Thursday evening, I was eating my last meal in the *kibbutz* dining hall, when my teacher approached and sat next to me. She started again to try to convince me not to leave the *kibbutz*, and assured me that I would not regret staying. I just listened and did not respond. After I finished eating, I went directly to my room and did not leave until morning. On the following Friday, I woke up very early, packed all my things in my suitcase, and I took the bus to Afula. The central station there was overcrowded with people, especially soldiers who wanted to return home from their military bases for the weekend vacation. Many of them wanted to save the ticket money and stood on crossroads to hitchhike. I got on the bus to Tel Aviv and during the whole trip, I was deep in thought.

I felt confused since I did not know if the decision I had just made was correct. We were in a new country where we had just arrived and a good decision had to be made regarding our future.

I started to compare our current situation to the one we left in Kabul. My heart was crying especially for my father who was a prominent personality in Kabul and other cities of Afghanistan. I was recalling his friendships with mayors and governors as well as his leadership in the Jewish community. The secretary of the Afghan king was a frequent guest in our home. Above all, he was the head of our family and nothing had been decided without his approval. Then he came to the Holy Land close to retirement age to wander with a shovel and dig holes just like the Jewish people when they were slaves in Egypt. I could not accept this change because it was so painful. If he had been younger, I would have been more tolerant of this change. My father imagined various scenarios of his arrival to the new country, but he could not foresee the real circumstances he was encountering in Israel.

I wondered if he would have done this move had he knew the consequences in advance. We left a state where everything was in abundance. It was a Diaspora indeed, but the Jews there adapted very well, especially in Kabul where the people are generous, noble, and respectful of others. I did not know of even one Jew there who worked as a plain laborer. Everyone of the Jewish community in Kabul made a living on a high level. The same thing was the case of Jews in Hirat.

My grandfather came to Israel and found work in the municipality of Tel Aviv sweeping the streets in the central bus station. Two uncles of mine were street peddlers selling teacups to passersby. My brother-in-law Levi and his brother were gardeners who worked in people's yards. My uncle Isachar was a porter who could not even afford to buy a truck and used a motorcycle with three wheels for moving. This was the situation we encountered in the land Israel, and on this trip home I realized it very clearly.

I arrived home and found only Dad there. He told me that the rest of the family had already traveled to visit Uncle Nafthali in Tel Aviv for the weekend. He remained to wait for me. He said, "I am happy that you have already arrived. Take a shower and let's join our family at Nafthali's house. All of them are waiting for us." I was happy that I returned and more happy that we would have the Shabbat together.

CHAPTER 18

After four weeks of intense Hebrew studies, I had finally returned home from the *kibbutz*. This short period was a nice experience in my life since I experienced living far from home independently, and besides learned the basics of the Hebrew language. I had become much more familiar with the new country and its people. Now I was standing at a crossroad of my life. I needed free time to think and to consult with others to decide what the best thing to do was. Everyone gave me a different suggestion and I didn't know who to listen to. The days passed and I had not yet done anything other than think. My brother-in-law Levi, who saw my situation, asked me to work with him in gardening in the meantime. I accepted his idea and worked with him during the summer taking care of the yards of homes.

I would soon be eighteen and expected to get drafted to the army. Until then, I decided to work for half the day, and the other half studying a profession. I registered in a vocational school in Tel Aviv called Montefyori with a major in electronics. The studies would start in October after the high holidays and would take place daily from four to seven p.m. Meanwhile, I worked in gardening and simultaneously looked for another job. Once a week, I went to the employment bureau to look for work but they had nothing to offer. I was told to return after the holidays.

We received a letter from our family in Afghanistan. My two sisters there and their husbands had already gotten their passports and visas to leave the country. We missed them very much and especially their little children, Dahlia and Yosef, who we had watched grow up. This letter gave us a lot of hope that within the upcoming year they would all be with us in Israel.

During the period of the high holidays of *Rosh Hashana*, *Yom Kippur*, and *Sukkos*, we were at Uncle Isachar's home and felt very comfortable there. It was a pleasure that all of our relatives and friends in Israel lived in two close neighborhoods; Shapira and Kiryat Shalom in Tel Aviv. We enjoyed sitting together with the children of the families and conversing in Hebrew. Right after the holidays, we returned home since Dad had to be back at his work.

When we returned home, I received a notification from the employment bureau telling me that a proper job was available for me. This was a state bureau which belonged to the ministry of labor and was in charge of supplying unemployed people with work. I went there and the officer of the bureau offered me work in Yahud in a factory called Relif that colored fabrics. I immediately accepted the offer and returned home. I asked my sister to accompany me to this plant which was in the industrial zone; five minutes walk from home. I met the manager there, a Hungarian Jew named Mordechai. He interviewed me and decided to hire me for work right away. Since I had to study radio electronics in the late afternoons in Tel Aviv, he agreed to keep me at work until two o' clock daily.

My boss was a nice person and instructed in my work. It was manual labor which was a bit physically difficult but I survived it. My purpose was to keep this job until I finished my studies. Thus, I could financially help my parents who needed the money for the household.

The year of 1967 passed by and at the beginning of 1968, I got the first summons to the army. My brother Amnon read it carefully and explained to me that since I was a student, I could ask for a postponement to the draft. I traveled to the recruitment office which was in the city of Petach Tikva and told the officer there what my brother instructed me. The officer, who was a religious man, approved to let me continue my studies for another year to 1969. So I continued to work and study simultaneously. It was very difficult for me since for half the day I worked physically hard and the other half I struggled with my studies. I wrote down everything my teachers lectured and finally, I succeeded to build a radio. Complaints of my hard situation would not have helped; I had to complete these two tasks that I had taken upon myself.

After the Six Day War, there was plenty of non-skilled work to be offered. My friends of my age in the neighborhood worked only in hard physical labor and earned good money. Much work was offered in military and civil construction. One friend asked me if I wanted work for good money. I accepted but only for half a day. He said that it was fine since the work he wanted to offer me paid by the hour. The work was in the industrial zone of Petach Tikva and consisted of loading heavy crates of olives onto trucks.

I left home at seven o' clock in the morning and since the bus only arrived on the hour, I reached the plant at 8:30. The owner welcomed me with a severe look and said, "We start work at 7:30. You are late. Go home and come back tomorrow on time." I responded, "But I am already here and can work five to six hours now." "No!" he insisted. "You should be on time." He did not give me a chance and I returned home shamefully. The next morning, I left home at six o' clock and arrived at work on time. I worked for the whole day and it was enough for me. It was work for horses and not human beings. The next day, I returned to my previous work at the Relif plant near my home.

We received a letter from my sisters Bracha and Tzippora in Kabul informing us that very soon they would leave the country and join us in Israel. They wrote to us that Afghanistan was not the same state that it was before and all the Jews were going to leave the country. The places they preferred to move were America, Israel, and England. If everything would go according to plan they would arrive in Israel within a month. I received another letter from my friend Yusuf in Afghanistan in which he wrote that he missed me since he did not have any other friends. He wrote that there was a high possibility that he would travel to India to continue his education.

Naturally, we were happy that the whole family was going to be reunited as we were in Kabul. We counted the days until the arrival. The only one who had extreme difficulty adjusting to the new country was my father. He was working very hard but did not complain. My mother tried to encourage him, but he was suffering. One day, she went out for shopping and to her surprise, she saw Dad during his work. The sight shocked her. Half of Dad's body was underground while he was digging and above him two managers were standing giving him instructions. They said, "Mr. Cohen. Dig deeper!" Tears fell from Mom's eyes and she approached the managers saying to them, "How could you do this to him?" They laughed at her and said, "It's okay, Ms. Cohen. We just protect him." My Dad, who did not like Mom's intervention, looked at her angrily as a hint for her to leave the place. At home, he explained to us that Israel is a tough country where it is very hard to make a living.

The days passed and one day in 1968, we received a telegram from Tehran, the capital of Iran. It notified us that my sisters Bracha and Tzippora and their families would arrive at Tel Aviv airport the next day. My cousin Ilana and I were sent by my parents to welcome them at the airport. We went there but did not know where to meet them. At that time, the terminal was not as big as it is today so we could look all over until we finally found them at an exit which was special for new immigrants. We saw them through the glass wall but could not get in. We waited for half an hour until they got out and the meeting was very emotional. My brother-in-law Tzvi, who was my teacher in the past and had a very close relationship with me, embraced me with tears of joy. The little children, Dahlia and Yossi, ran to me to hug and kiss me. I thought the separation of a year and half from the children would cause them to forget us, but I was wrong.

In the Jewish Agency at the airport, they were referred to an apartment in Tel Aviv in a neighborhood called Kfar Shalem. This was a good location with houses that had just recently been built, so they had better luck than we had when we first arrived. I helped them carry their suitcases to their new home and immediately went with them to our home in Yahud for a celebratory dinner.

They asked how we were doing and enquired very thoroughly to what was going on in Israel. We explained to them that here, everything is different compared to Afghanistan but with time they would get used to it. The mentality of the people here is different as well as the atmosphere and even the food. However, it does not take long for the youth to adapt. The only ones in the family who had difficulty adjusting were my parents, but they had no choice once they were here. Anyhow, they received encouragement from visiting friends from America and other countries abroad.

An old, good friend of Dad, *Mollah* Meir Shamash, and his brother Yehuda, came from America to tour the country. Dad met them at the Hilton hotel where they owned shares. They left Afghanistan in the 1950's and got rich in the United States. Their company handled import and export businesses mainly of precious gems and diamonds. As their company grew very fast, they opened offices in Japan, Hong Kong, Thailand, India, Italy, and Switzerland. The main office was in New York. They were very thankful and full of gratitude to Dad for the favors he granted them in the past. During this meeting, they offered him to send his son to work with them in the states, as they needed more personnel and they naturally preferred one of their acquaintances.

My father gave his consent on the spot. Dad returned home and excitedly told me about the conversation with his friends. He told me that this was a great idea and encouraged me to take the offer. My response was not so enthusiastic. I said to him, "Everything in this plan sounds good, but do not forget that we had just left Afghanistan two years ago to Israel, and it was not an easy until I got used to this new country and learned its language. To leave all this and to move again to another foreign country, means that I would have to start all over and I do not have the patience and strength for that. Besides, there is no way that the Israeli army will let me leave abroad at this time since in one year from now, I have to be enlisted."

Dad tried another way to convince me to accept the offer by saying, "Look around you. This country is constantly busy with wars and its military situation worsens. Every day, there are battles on the borders, especially with Egypt which is called a war of *Hatasha*. The Egyptian's strategy is to keep the border under attack daily, thus weakening the Israeli state. I do not see any good future for you in this state."

All the efforts of persuasion by Dad until now did not help since my mother supported me in my enthusiasm to get enlisted to the Israeli army regardless of the consequences. In 1969, I again received a summons from the army. This time, I had to pass very difficult exams while I still did not know what their plans for me were. During a whole week, I took written tests for five hours daily. The army wanted to test the level of my professional knowledge. At that time, I had already graduated the vocational school with a major in radio and electronics and received a diploma. At the end of the exams, they gave me the date of recruitment which was February fifth, 1970. Until my service began, I continued to work for a full day in Yahud near my home.

On Saturdays, our whole family who were united in Israel used to gather together. Particularly on *Motzei Shabbat* our uncles and aunts joined us with musical instruments. Asher led the 'family orchestra' by playing on a mandolin which he learned in Kabul and others played drums. The whole family accompanied them in song. All of the songs were from Afghanistan and they reminded us of the good days and adventures we had there. These were evenings full of nostalgia.

When I compared those wonderful memories from our old country to the hardship the newcomers encounter in Israel, I was not encouraged. Everything in Israel was difficult to achieve, but it should not have been a surprise for us since our sages taught us that three things are achieved only after suffering. First, the land of Israel, then knowledge and understanding of the Holy Torah, and finally, the merit for the Next World.

In that year, my brother Amnon came from Tel Aviv with the bad news that grandma Sarah had been hospitalized for a severe sickness. Mom immediately wanted to leave everything to go and visit her, but we kept her back since the day after, Dad could have joined her after his work. We did not have a telephone since at that time, it was a very expensive item in Israel and very few had one.

Dad had a phone at his work, which was now in a factory for carpentry glue near our home. This was his first steady work in Israel in which he was a supervisor for laborers. On the day they planned to visit grandma, he received a phone call at work that she had passed away. Dad immediately left work and came home. Since he had arrived home earlier, Mom understood that the worst had happened.

All of us traveled to Uncle Isachar's home in Tel Aviv. All of the family members were already there and everybody comforted grandpa Yehuda who was sitting outside on the porch. The funeral was managed very respectfully and the coffin was carried to Jerusalem to be buried at Har Hamenuchot, a renowned cemetery in the Holy City. Mom was comforted by two things, that her mother did not suffer a lot before passing away, and that she arrived in the land of Israel before her mother's final departure.

We had another good reason to be thankful to G-d that we arrived in the Holy Land despite the difficulties and the wars. We received news from Afghanistan that a revolution was taking place there. We actually expected this change for a long time but did not know the exact time when it would happen. The king went for a visit to Italy and his nephew, Davod Chan took advantage of his absence from the country and started a revolution. He declared the state as a republic and appointed himself as the president.

The Russians, who shared a border with the country, were very involved in Afghanistan towards the end of the king's regime. Russians could be seen everywhere. They were consultants for the Afghan army as well as the local infrastructure. Factories were established by them and many employees were Russians. The Afghan military was armed with Russian equipment. The king was interested in establishing close relations with America and the Western world but under the Russians' influence, he could not implement his wishes. The assistance of the West to this poor country was restricted by the Russians.

The nephew of the king, who once served as prime minister, wanted a more modern country and it was expected that he would reach power some day, but we did not expect a revolution from him. Anyhow, the Russians did not want to give up their strong influence over the country and thus they invaded it. They killed all of their opponents and established a pro-Soviet regime. Since then, Afghanistan turned into a country of terror. Only then, we realized the good fortune that the Jews of Afghanistan had previously. All of them left their country with their possessions via Pakistan and Iran before the chaos in the country started.

During the summer of 1969, the conflicts with Egypt got worse. There were shellings and casualties every day on the Israeli-Egyptian border. Israel suffered at least one dead soldier a day. One Sunday, I traveled to Tel Aviv to make arrangements and returned home by bus in the afternoon. The bus drivers would turn the radio up during the news broadcast. On that ride, the radio broadcaster announced that an Israeli soldier, a twenty-four year old, from the city of Kfar Saba, was killed near the Suez Canal. When I heard the name of this casualty, I was shocked and in a state of disbelief. The soldier was Moshe Kashi, a good friend of mine. He was a decent and pleasant person. We grew up together in the same neighborhood and studied in the same school.

When I returned home, my first question to my mother was if she heard about the bad news. She said, "Yes, it's true. The fallen soldier was your friend Moshe." Then Mom told me that that evening we were invited for dinner by my sister Bracha and her husband Tzvi, and that we might hear more details about the incident. All of us traveled there but they knew no more than we did. Besides, they also did not have a phone to call the stricken family.

The next day, my brother-in-law Asher, who was a very close friend of Moshe, traveled to his parents' home in Tel Aviv. He found out that all of Moshe's close relatives were sitting *Shiva* at his uncle's home. Asher returned and passed over this information on to us. We traveled there to comfort the mourners. We met a family in shock and the parents were barely conscious. We could not talk with any of them except with Moshe's younger brother Eliyaju. He was the only one who could function and gave us details of what happened. Moshe was killed in the line of duty by an Egyptian sniper right on the eastern banks of the Suez Canal.

This incident was one more tragedy for the people of Israel but an especially big tragedy for the Afghan Jewish community. A fallen Israeli soldier was a daily routine and the psychological effect on the people was worse than in a regular war, since it lasted for a long period with no end in sight. Even though, the Egyptians suffered many casualties themselves, including property damages, by the Israeli army, they did not give up. They shelled the Israeli side every day and their snipers were always ready to shoot any Israeli to come in sight. The Soviet Union, who supported Egypt, was interested in the continuation of the conflict with Israel since they could sell more military equipment. The Israeli people got exhausted from this type of long-lasting war and prayed for its quick conclusion. The time for my military recruitment drew closer and me and my family prayed that until then there would be peace between Israel and its Arab neighbors.

My brothers-in-laws Tzvi and Asher suggested that I take a vacation from work for a few days to travel with them to the holy places of Jerusalem. We left on a Thursday morning by bus to Jerusalem. We stayed in the city was at the home of Tzvi's sister, who had lived there since 1950. On the first day of our tour, we traveled the whole ancient city of Jerusalem and afterwards, went to the city of Hebron to visit the tombs of our patriarchs, Abraham, Isaac, and Jacob who were buried in the Machpellah Cave. We then traveled to Shchem (Nablus), another nearby ancient city and visited the tomb of Joseph the Righteous, son of our Patriarch Jacob.

At these very holy places for the Jewish people, we prayed for the peace of the state of Israel as well for the whole world. In the evening we returned to Jerusalem. On Friday, we were wandering through the Arab market of the ancient part of Jerusalem. For Shabbat, we were guests at the home of Tzvi's sister and on Saturday evening, we visited *Schunat Ha'Bucharim*. On Sunday, we spent the day in the modern part of Jerusalem and visited the *Knesset*, governmental offices, and the national Israeli museum. In the evening, we returned to Tel Aviv.

To summarize this tour, I would say that it was a very exciting and enjoyable one. We felt an atmosphere of holiness everywhere we visited and the ancient sights touched our hearts. Crowds of people, Jews and non-Jews, filled the narrow streets and alleys of old Jerusalem and its surroundings. They came from all over the world to tour these places. I was very convinced that these ancient cities of Jerusalem, Hebron, and Shchem were now in the right hands after being under foreign rule for thousands of years.

The state of Israel as well as private entrepreneurs started to invest and build all over the West Bank of Jordan, which had just recently returned to the Jewish people. New roads and sidewalks were built there for the convenience of all the tourists that came from all over the world.

The relations between Arabs and Jews in the united city of Jerusalem were very good and the people who traveled there spent a lot of money. The Arab merchants welcomed them very warmly and their business flourished. I was very excited to see the pleasant relations between Arabs and Israelis and could not understand why there were such bitter wars between Israel and the Arab countries. Why could they not get along? Jews had succeeded for hundreds of years living in peace among Muslims in their countries like we did in Afghanistan. Why is the Holy Land always in war? Is there anything more beautiful and good than peace?

My time of recruitment was very close and my family was getting anxious. They said to me, "Michael, be cautious and don't volunteer for combat. Tell the military authorities that you are from a family with many children so they can let you serve closer to home." In Israel such a soldier is nicknamed *'jobnik'*. I just listened to what my relatives said and did not respond. I thought to myself, "Either I am a real soldier or I will have no interest in the military." From time to time, I came to hang out in Tel Aviv and always visited my grandpa Yehuda who was feeling lonely after grandma's death. He moved to live with his son, uncle Nafthali. I enjoyed staying with him very much, since he was always calm and never nervous or angry. He always gave me lessons in morality and constantly blessed me.

A week before my recruitment, I came to grandpa to bid him farewell. He was not in his room and my aunt told me that he had left home to the synagogue in the early morning and had on that day *Ta'anit Dibur*. A group of observant Jews were sitting together in the synagogue studying the Holy Scriptures and did not speak a word. In the Jewish religion, this is more appreciated then fasting in order to express piety. I went over to find him studying and he would not talk to me. I whispered into his ear, "Grandpa, next week I am going to be drafted." He looked at me, put his hands on my head, and blessed me. With a smile on his face, he bid me farewell. This was not the proper time to come and visit him.

The last week before my recruitment, I was on vacation. My mother invited the whole family to our home and almost all of them came. The purpose of this gathering was to celebrate the 15th of the month of *Shvat,* the New Year's for the trees, and is a holiday of gratitude to G-d for the fruit that he has bestowed on us. The plan was to celebrate as we would do in Kabul by offering the guests a huge variety of dried fruit. Unfortunately, we could not find more than four types in Israel, compared to the over twenty types of dried fruit in Kabul.

In Afghanistan, this was a big festival in which each Jewish family would send a basket of fruit as a present to another family and the *Yeshiva* students did the same for their teachers. This was the custom unlike in our land of Israel where the holidays and festivities were celebrated less strictly. Anyhow, we were getting used to this new situation and gradually started to behave like Israeli's.

CHAPTER 19

The day of my recruitment, February fifth, 1970, finally arrived. I woke up in the early morning to find Mom already awake and very excited. She prepared food for me for the trip. I took a backpack with me and Mom accompanied me down the stairs with a bottle of water which she poured behind my footsteps for good luck. She blessed me by saying, "*Choda Hamrat,*" and I left. I arrived in the city of Petach Tikva at the recruitment office. Crowds of youths were waiting outside to be called by name and board buses. My bus left at 8:30 a.m. to the first military base at Tel Ha'Shomer.

We got off at a huge open space. An officer welcomed us with a scream to stand into groups of three. From then on, we were under military discipline and each instruction was an order that had to be fulfilled. We marched after him and one by one, we entered the military office to get our dog-tags with a personal military number. This was recruitment number 209 of the Israeli armed forces since the state was established. We received immunization shots, military uniforms, and a huge duffel bag with clothing for various purposes in it. The officer then guided us to tents where we spent the night. Each tent had contained ten beds.

The morning after, we met the assignment officer who decided where to refer each one of us. My turn came and he looked into my personal file and asked me if I wanted to serve in the paratroopers or the armored forces. I answered that I wanted to be a combat fighter and perhaps I could serve in the navy since I liked the ocean. He determined, "You are suitable to be a soldier in the armored forces.

Good luck." Then I had to leave. Many soldiers did not want to serve far from their homes as they did not want to be fighters. The bottom line was what this assignment officer determined. There was no way to change his decision.

A military truck arrived at the spot and a group of twenty newly enlisted soldiers, including me, got on board and the accompanying officer told us that we were going to a base of the armored forces in the Gaza Strip of the Sinai Peninsula. Some of the soldiers could not believe their destination since they asked to become paratroopers and they complained during the whole trip. I was sitting quietly since I did not fully understand the differences between the branches of the military. As a true believer in G-d, I had faith that He would guide me to the proper unit where I could serve the country in the best way. I had the feeling that serving in the armored forces would be a proper platform for my patriotism. I knew that this branch of the army contributed the most after the air force to the Israeli victory in the Six Day War.

The Israeli tanks rolled throughout the whole Sinai Peninsula very quickly and efficiently and reached the Suez Canal. My military base was in a large area and very organized. There were plenty of small residences for the soldiers with ten beds in each room. We had a beautiful common dining room and gym hall. There was a synagogue in the base with a chaplain who took care of the spiritual needs of the soldiers.

The discipline in the base was very strict since Yulius, the officer in charge of it, was a very tough person. He was tall, well-built and he scared every soldier who got close to him. Since the first day in the base, we experienced the harsh reality of life in the military. Every order given by our staff sergeant had to be fulfilled immediately and quickly. He drove us crazy since the beginning. To go to the dining room, we had to march in groups of three. One of the soldiers who did not like this treatment shouted, "The sergeant is a son of a bitch!" The sergeant heard this and demanded to know who cursed him. There was no answer and he immediately responded by punishing all of us.

For a quarter of an hour, we had to crawl up and down a hill and stand up and sit back down according to his orders until he finally got bored and released us to the dining room for dinner. We actually expected things like this and worse to happen, but we accepted all of this in good faith. Most of the soldiers that got recruited with me were from agricultural settlements. They were stronger, had more patience, and followed all orders strictly. Nevertheless, they always had unnecessary questions. Twice a day, we ate meat that was served in small, personal containers called *mestings*. Before every meal, the sergeant would organize us into groups of three while we held our *mestings* and we marched to the dining room.

This military base had many soldiers since it dealt with basic training. The schedule of the day was very busy. We woke up at 6:00 a.m. and started our day by running with a sports coach. Afterwards, anyone who wanted to pray in a *Minyan* could do it in the synagogue. Breakfast was served at 7:30 for about forty minutes. At 9:00, the theoretical military training began with lectures by high ranked officers for about an hour. We then had a short break, and afterwards practical military training with submachine guns until 12:30 when we had lunch. The training resumed at 1:30 and lasted until 5:30 in the afternoon when we ate dinner.

At night, we had an evening run and then each soldier had to serve a shift as a night guard for two hours. This intensive basic training would last for three months since the army needed new fighters to join the front lines at the Suez Canal. We had no free time whatsoever. We did not sleep for more than five hours and all our time was dedicated to training and live fire exercises. On Saturdays, we had the day off from training, but still each soldier had to serve a work shift in the kitchen or as a guard.

Once in three weeks, we were allowed to go home for the weekend. Before leaving home, there was a thorough inspection in the rooms to make sure everything was clean and organized. The first time I left the base, I had to be dressed in a special dress uniform with my personal submachine gun, which was clean and shiny. Happiness befell on us when we saw the public buses entering the base to pick us up. I boarded the bus as a proud soldier who served in the Israeli armed forces. Back at home, my family did not know where I was or when I was coming back. In the base, there were no public phones or mailboxes. I got off the bus in the center of Tel Aviv where many soldiers were waiting at the *Trampiada.* Almost every private car owner would stop to pick up a soldier.

This was the time after the Six Day War when the soldiers were the pride of the country and they were respected by everyone. Everywhere in public, soldiers were given priority. They did not have to stand in lines. I arrived at Yahud in my military uniform with the submachine gun under my arm and marched to my home in long strides. Children of the community, with my little brother Abraham among them, were playing soccer on the street. This was a weekly custom for them before the entrance of the Shabbat. When they saw me dressed as a soldier, they got very excited and observed me from top to bottom. The kids followed me until I entered my home.

The table was already prepared for the upcoming Shabbat and my parents and siblings were surprised and happy to see me. They immediately asked me many types of questions, with seemingly no end in sight. Where do you serve? Which faction of the army? Is it difficult? And more. I was exhausted and all I wanted was to take a shower and rest. My whole body and head were dirty from the dust of the Sinai Desert which got sticky from my sweat. The physical effort from the training was tremendous and therefore, I had sweated a lot, but my muscles were stronger and more prominent.

As usual, Mom gave me her undivided attention and asked if I was hungry. I told her that all I wanted was to wash myself and rest. I knew that on this Shabbat Eve, I would have a lot to eat from my mother's delicious cooking. However, the food at the military base was not bad at all since professional cooks were employed there. I took a shower while my mother put all the dirty laundry that I brought with me into the washing machine.

Afterwards, I sat for a little while with my parents to answer their many questions. They did not understand so well since military life was not familiar to them. They repeatedly asked the same questions.

The most important thing for them to know was that I did not serve on the front line. "No! No!" I answered. "I serve near the Israeli city of Ashkelon and undergo basic training there." The only one who understood the function of the armored forces and where they would be assigned to was my older brother Amnon; but he kept this knowledge to himself. Finally, I was able to rest for a few hours.

Just before the entrance of the Shabbat, I got out of bed and dressed myself in my Shabbat clothes. For the first time in my life, these clothes seemed strange to me. However, they were light and pleasant on my body. The men of the family went to the synagogue near our home to pray and then returned for the Shabbat celebratory meal. Mom had prepared this meal and as usual, it was delicious. After dessert, I immediately entered my room, fell onto the bed, and slept like a baby until morning. Because I was so exhausted from the intensive training, I did not get out of bed for the entire Shabbat until the afternoon, when some friends visited my home after hearing of my arrival.

These friends were about to be recruited themselves very soon and naturally, had many questions to ask. How is it in the army? Is it difficult? I had to answer them and told them that the experience was relative. It depends how you look at your service. If you take it cool and easy, it is really not difficult. Otherwise, the time passes very slowly and you would be nervous and under stress during your entire service. They observed my submachine gun and started to play with it. I warned them not to touch it and assured them that when their time would come, they would also get guns. I thought to myself, "May G-d bring peace very soon to the land of Israel so my friends would not need to experience what I have." They left my home and in the evening my family and I traveled as usual to Tel Aviv to visit our close relatives there.

I woke up early Sunday morning to prepare myself to leave to the military base. Again, Mom followed me with the bottle of water to pour it behind my footsteps. The bus route to the base started its journey from a gathering point of soldiers near the Cinerama of Tel Aviv. It was a three hour ride with a temporary stop at the city of

Ashkelon. We arrived at the base in the late afternoon fresh and calm from our short vacation. Now the really difficult training was about to start. This new course of training would last for four to six weeks and take place entirely outdoors in an area thirty kilometers away from the base.

We went there by foot carrying all our equipment and ammunition on our backs. It was the month of March and the beginning of spring and the weather was very pleasant. We marched for six hours until we reached our destination. It was in the middle of the desert and no buildings were in sight; just heaps of sand. Every two soldiers set up their own small tent.

Fortunately, the soldier who was assigned to be with me in my tent was a friend of mine, who lived in the same room with me back in the base. His name was Shmuel and he was a very quiet, slow, and indifferent person. I had to do all the work for him. We set up the tent and put in all the personal equipment. It had become dark and we turned on flashlights.

Each soldier waited for the commander's call to get out for training. The call did come because heavy rain had started to fall. We became very happy and crawled into our sleeping bags. As we had been very tired from the long march, we all quickly fell asleep. We woke up very fresh in the morning to see the sun shining. The training started with morning exercises and then we ate a good breakfast. Afterwards, we listened to a sequence of lectures explaining in detail what we were going to do that day.

The plan was to mimic a scenario of real combat. We were wearing helmets and carrying weapons along with stretchers that had soldiers on them pretending to be wounded. This training was very difficult. We also practiced raiding a location we had to conquer using live ammunition. At night, we marched under the guidance of the stars about ten kilometers from our tents and back. We slept just three or four hours at night and the same routine repeated itself every day for four weeks in the open field. On the last night, the whole group followed the commander who used a flashlight in the darkness. We went to a *kibbutz* called Nir Yitzhak to take a needed shower. We were given half an hour for this shower and then we immediately returned exhausted to our tents. The morning after, we received an order to dismantle the tents and clean the area. We then returned by foot to our military base.

For four weeks, I was away from home and still did not know when we would get a break. We only received hints that maybe on the fifth or sixth week we would be able to go home for the weekend. The staff sergeant explained to us that if we would pass the commander's inspection, we would be entitled to this break. He told us that next week on the midnight of Wednesday, the commander of the battalion would visit each unit of the base and check each and every soldier – his weapon, his room, his bed, and even his shoes and underwear. He would thoroughly check if each item was clean and shiny and if the soldier was neat and shaved. Anyone who failed this inspection would not go home for the weekend. We received this news with a shock and immediately started to clean everything.

The expected day arrived and midnight had passed with no sign of the commander in the area. It was almost sunrise when the commander of the base arrived and told us that the battalion commander would not come, since he was called to an unexpected meeting at the chief military headquarters. Nevertheless, the base commander was willing to the inspection himself. We were relieved at this announcement. He approached me and asked to see my weapon. He looked into the barrel and said, "Congratulations. It is very clean." After finishing the inspection, he said to all of us that we had all passed excellently and he was proud of us. He wished us a pleasant vacation.

It was Thursday morning at 8:00 a.m. when the buses entered the base. The staff sergeant ordered us to stand in groups of three in our dress uniforms and with our weapons under our arms. Like this, we boarded the bus. All of us were very happy to return home after six weeks of intense training. I had my duffel bag with me full of dirty clothes to take home and clean. I felt very calm on the bus just from thinking that now I would have Friday and Saturday to rest.

We were just a week away from *Pesach* and at home, the

preparation for the holiday was at its peak. Mom was disappointed to hear that I had to go back to the base on Sunday. Nevertheless, she asked me to stay at home for the holiday and she was sure that I would not be punished. She said, "Just tell your officers that you have elderly parents who were lonely and needed you for the holiday." I answered, "Yes Mom. I wish it could be so. If everyone would use this excuse, who would protect the state?" For the few hours until Shabbat started, I went with some friends of mine to the beach. I looked great with prominent muscles and I felt very healthy and strong. This was the accumulated result of all the exercises and training in the army. This enhanced my self-confidence to physically beat any opponent as the army intended. I also had a deep tan from the sun rays at the Sinai Desert.

From Shabbat eve to the whole day after, I enjoyed Mom's special Shabbat meals and spent the rest of the time in bed. I had to relax my body and soul.

Again, on Sunday in the early morning, I left home wishing my family a happy holiday since I would not see them until after *Pesach*. The bus arrived at the base in the late afternoon. An important announcement welcomed us that our whole group was going to soon move to the front line at the Suez Canal. It was explained to us that we would replace reservists to guard a missile base.

These reservists were older and they received a break to spend the upcoming *Pesach* holiday with their families at home. We were all very excited and asked ourselves if we were prepared for this. We did not know exactly what we had to do there, since the explanations were very vague. All we were told was that we would serve guard duty during the week of the *Pesach* holiday until these reservists would come back from their vacation. We accepted all of this with no questions since we expected a great adventure.

A day before the eve of *Pesach*, a group of twelve soldiers from our base including myself, traveled three hours to this missile base. There were no buildings on the base except for bunkers with guard-posts. In the guard-posts, only the helmets of the soldiers could be seen. However, they were always ready with anti-aircraft artillery.

In this part of Sinai, the color of the sand was close to yellow and the sun was very strong. At the beginning, we were blinded by the sunlight but we got used to it after a while. The commander of this facility was a young and talkative man. He gave us detailed instruction on what should be done.

The next day, the commander divided us into two groups for the evening. One group of soldiers would be at the guard-post while the other group would perform the *seder* in the bunkers. When the *seder* was finished, the groups would switch places. On the eve of *Pesach*, Jews around the world sit together with their families to eat a celebratory meal accompanied with a recital of the story of the Exodus from Egypt.

On this particular eve, I had to celebrate far from my family in this weird place under the ground. We naturally accepted the commander's decision when suddenly, we heard someone scream, "Catch it! Catch it! It escaped from me!" I did not know what it was but then I saw the kosher supervisor running after a gas burner. This supervisor came especially to the base to keep its kitchen Kosher for Passover. This would be done with a manually-operated gas burner that emits huge flames. While he was operating this device outside, it went out of control and moved by itself from the strength of the escaping gas.

It danced all over the base while all of us chased after it. The commander shouted, "Goddamn! Catch it before it reaches the gas tanks!" I ran crazily after it. Nearby were sacks filled with sand. I emptied one of them and wrapped my hand with it. I took another full sack and threw it at the device which stopped moving. I then approached it and turned it off. The commander looked at me in surprise and said, "You just prevented a possible major disaster. I am giving you two points." I answered, "What? Two points? For such an action, I deserve a vacation and a promotion."

Then the commander said to me, "I agree that you deserve such rewards but I cannot grant them now. Anyhow, I will give you a recommendation." Right after this talk with me, he approached the kosher supervisor and chewed him out. At sundown, we celebrated the first night of *Pesach* in the bunkers. The week of the *Pesach* holiday was a time of relaxation for us. We did not have any training, marches, or early awakenings. All we did during that week was four hours of daily sitting in the guard-post. Only in the event of an early warning, we stood guard for longer periods.

After that week, we returned to our base to again endure difficult training for one more month. During this month, we had to participate in the yearly three-day march of Jerusalem. This march would take place in Jerusalem and its neighboring communities. The purpose was to commemorate the day of the reunification of the city which occurred during the Six Day War in 1967. Another event which was planned for us was the oath ceremony on the Massada Mountain. This place near the Dead Sea was considered sacred by the Israeli army. Around two-thousand years ago, the Jews rebelled against the Roman Empire and fought until the bitter end at the top of the mountain, where they took their own lives in the final moments. At the conclusion of the month when basic training would end, a party was planned for all the soldiers and their families. I sent invitations to my parents and to my older brother to attend and they confirmed that they would be there. I was very happy.

Before the "Day of Jerusalem", we left the base on buses to the holy city. Each one of us carried two handbags; one with the daily uniform and one with the dress uniform. We arrived at an open space in the city and received an order that every two soldiers should build a tent for themselves. We did this very quickly. All of the open spaces in the city were covered with tents by soldiers of all factions of the army who came from all over. Military trucks delivered food to these temporary camps. Portable lavatories and kitchens were set up. Tables and chairs were placed for dining. We were ready to take part in the three day march that was to take place on the next day. The whole city looked like a military carnival. Flags from all of the military factions decorated Jerusalem.

We woke up in the early morning for the first day of the march. We had no helmets or bags to carry and were able to move easily. We just marched and sang and had high morale. From time to time, we shouted, "The Israeli Army is the best!" An officer led our group and a staff sergeant marched parallel to us to make sure we were on our best behavior. We passed forests, marched over hills, and descended into valleys. Small Arab villages were on the path of our march, and some Arabs were in the field with their livestock. They looked at us in astonishment as we marched quickly together in song. The roads were covered with thousands of marching soldiers.

On the first day, we marched for fifteen kilometers. Some of my friends had to drop out from the march because they could not stand the pain of foot blisters. Thank G-d that I was in excellent shape and easily completed the march. We returned to our camp in the afternoon for dinner. Afterwards, we entered our tents, took off our shoes, and realized that our feet were also covered with blisters. We received talcum powder to take care of this. We then rested until the evening.

After dinner, all of us were invited to see a show in a nearby amphitheatre. Right before the show started, the military mailman called me and said, "Michael, I have a letter for you from Tel Aviv." I was wondering who would have sent me a letter and I opened the envelope immediately. It said, "Hi Michael. Your Dad gave me your military address so I can send you a letter. If it is not too difficult for you and you do not object, I would like to be in correspondence with you. I know that your life now is not easy and you have passed difficult training. Besides, you are very far from home somewhere in the Sinai Desert. We wish that you take it easy and not break down. We are proud of you. Good luck, Miriam." This was a great surprise for me. Miriam is the daughter of my uncle Nafthali and I really appreciated her initiation. I felt that I had to answer her right away. As the show started, I was sitting and writing on a postcard on my knee. It was a short letter in which I wrote how happy I was to receive her letter and asked her to pass over my regards to the rest of the family. I mentioned that I was now on the three day march in Jerusalem and was expecting more letters from her. I gave my postcard back to the mailman.

The show was pleasant. The Israeli renowned singer couple Ilan and Ilanit appeared and all the soldiers had a good time. The amphitheatre was half empty since many soldiers preferred to take care of their injured feet and rest in bed. We returned to our tents at 10:00 p.m. and immediately fell asleep.

Towards the second day of the march, we woke up very fresh but in pain from the blisters. It was difficult to put our shoes on. Anyhow, I put a lot of powder in my socks and was prepared to march again. This time our group was smaller since several of the soldiers dropped off. The route of this day was more difficult since it was on a mountainous area and there were more kilometers to pass. Even though it was not an easy task, I enjoyed it. During the march, my group sang many types of Israeli songs and that is how I learned these songs, besides the ones I learned during my military service. That evening, the soldiers went to the amphitheatre again to watch a performance by military bands and civilian artists. I had never seen them before and all of us were rolling from laughter. The purpose of these performances was to enhance our morale to the maximum and this was achieved indeed.

In the morning of the third and final day of the march, I woke up with terrible pain from the blisters. It was difficult for me to take a step. I could not miss this day of the march since at the end of it, each soldier would receive a medal and certificate that proves he attended and finished the whole march. I again put lots of powder into my socks and prepared myself as if for a battle. I did my best to ignore the pain. Our commander encouraged all of us by promising that today would be the most exciting day of the march. The whole route would be in the city of Jerusalem. From the minute we left the camp, we did not stop singing and shouting slogans in favor of the Israeli army and especially the armored forces. We marched through the streets of Jerusalem while crowds stood on the sidewalks applauding as we passed by. The roads of Jerusalem were full of marching soldiers. The leader of each group carried a flag that represented his unit and the atmosphere of the city was like a carnival. In the afternoon, we returned to our camp, ate lunch, and went to the amphitheatre. There, a ceremony took place to mark the end of the three-day march in which medals and certificates were given to everyone in attendance. In the late afternoon, buses arrived to take us back to the military base in the Gaza Strip.

While returning to the base, we thought that we would get a vacation for a few days after the hard time we had during the march. To our surprise, the plan of our officers was quite different. A new group of recruits were scheduled to arrive soon to the base and we had to finish our basic training before they would come. We went to dinner and were notified that when we finished, all of us had to assemble in the gathering hall to attend a lecture of the history of Israel. A prominent lecturer would address us and no one was allowed to miss this event. We were extremely tired but we naturally preferred this to an evening run or any other type of physical training. The officer introduced the lecturer and warned us not to dare fall asleep. He added that anyone who would fall asleep would be punished with four hours of guard duty that night. The lecture was very boring and we could hardly keep ourselves awake. Suddenly, we saw the lecturer laugh while he was looking at the officer who had fallen asleep on his chair. It was very inappropriate for us to laugh, since he too had endured the march with the rest of us. After a short while, the officer felt that we had noticed him napping and he roused himself. However, the lecturer shortened his speech and the day ended peacefully.

Towards the end of basic training, all that was left for us was to attend an oath ceremony on the top of the Massada Mountain as well as a day of sports competition. Afterwards, we would celebrate our graduation from the three months of harsh training. Usually, this type of training takes six months, but because of the constant military clashes with the Egyptians, we had to pass through quicker and more intensively. We were specially trained and prepared to serve as combat troopers near the Suez Canal, either in tanks or as infantry, but naturally, some would be referred to do desk work.

Even though we had just a week left in the base, we received a weekend vacation to go home. As usual, I brought with me my dirty laundry to clean at home and I changed from my military to nice civilian clothes for Shabbat. I told my family about my adventures during the three day march in Jerusalem and reassured the attendance of my parents and Amnon in the upcoming graduation ceremony. On Sunday, we returned to the base and were notified that anyone who would fail in the upcoming sports competition would have to stay for an extra three months of basic training.

On the designated day, we traveled on public buses to Massada. Everyone was wearing their dress uniforms and carrying weapons. We ascended the mountain and received Bibles. A few commanders gave a speech and then we swore our oaths to the Israeli army to be loyal soldiers to the army and to the state of Israel. The whole ceremony was short but we had to wait a long time for it to start. When it was over, we immediately returned to the buses and to our base.

The next day was the sports competition. We felt more confident and talked freely with our commanders like friends. The competitions included a three kilometer run, obstacles courses, tugging rope, and a soccer game. I participated in the three kilometer run over a track that surrounded the whole base. Some soldiers failed to make it but even though I was among the last to finish, I finished. After these competitions, our training was finally complete.

At that time, every soldier was nervous regarding where he would be assigned for the remaining 33 months of military service. The commander of the base called each and every soldier to a personal conversation of about five minutes, during which he explained to the soldier where he would be assigned. When it was my turn, I entered the commander's office in excitement. He told me to sit down and said, "I am going to assign you to one of the most important battalions of the army which is number 184 located on the front lines. This battalion is composed of tanks and armored vehicles and you will be proud to be there. Good luck." This determination from the commander was final and no one could dispute it. I rose from my seat, saluted him, and left the office.

Friends of mine were waiting outside for their turn. When they saw me, they immediately asked, "Where to?" I answered, "Tank battalion number 184." They said that this battalion was located right near the eastern banks of the Canal. Some of my friends tried to appeal their assignments but without success.

On our last day in the base, I waited for my parents and my brother Amnon to arrive for my graduation ceremony. All the rest of the soldiers were also waiting for their parents and relatives to come. We wore dress uniform with berets on our heads and our weapons on our arms. We were standing on the parade grounds in groups of three when the buses with the guests arrived. All the soldiers were excited and were looking for their families. Suddenly, I recognized my brother who was standing near the driver of one of the buses and looking for me. I could not leave the group but knew that we would finally meet. We marched in groups of three lead by the commander towards the amphitheatre.

All of the guests took their seats and the ceremony started. The commander of the base delivered a speech and then each one of the soldiers got onto the stage and received his graduation certificate. The commander wished each soldier personally a pleasant military experience. At the end of this ceremony, the commander declared that we were all dismissed for the day and we all ran to greet our families. As usual, Mom blessed me and comforted me by saying, "Do not worry. Everything will be alright with you despite the difficulties." Dad was silent and only asked when I was coming home. I answered him that as of that moment, I did not know and I added, "Tomorrow, we are leaving to our new base in Birt Madeh which is in the center of the Sinai Desert. We will have advanced training there for two and a half months and then afterwards, we would move on to the front line at the Suez Canal opposite the Egyptian city of Ismaelyia to protect the fortress there."

At the base Amnon met an officer who was a friend of his from Tel Aviv and consulted with him about me. He told my brother that my final destination is a good place to serve with a very strong and active battalion. He himself had served there before. We bid farewell to our families as they left the base back home. We left the next day.

CHAPTER 20

On Wednesday morning our busses traveled to our new base in the desert. The residences were huge tents with ten beds in each. I chose a tent in the corner of the camp with two other soldiers I knew. One of them was of Iraqi origin and his name was Menashe. He was from the town of Or Yehuda. The other soldier was named Shemesh and he was from the city of Be'er Sheva. I hadn't met the rest of the soldiers in my tent yet. These new soldiers were stressed and nervous. I couldn't understand their behavior, since we all had to be together for many months; and in my view, it would have been much nicer if we could have gotten along. We prepared our beds and went to the command center to get equipment, work-clothes, and weapons. Each soldier had to sign when he obtained these items.

We went to dinner right after and got acquainted with the base commander there. In the dining room, everyone was stressed since we did not know what was ahead of us. We knew in general that someday we would move to the front line at the banks of the Suez Canal, and feared this. There were daily clashes between the Israeli and the Egyptian armies who stood at opposite sides of the canal. Israel used its powerful air force during the day and the Egyptians responded with nonstop shelling at night. This was what seemed like an endless war, which had already endured for almost three years. Each week, the Israeli forces suffered deaths and casualties. It looked like a game where the loser would be the first to give up. The Egyptians were eager to avenge their defeat in the 1967 war and did not want to accept the reality that they had no chance against the Israeli army which was well-equipped, better trained, and had higher morale.

This was the month of May and the weather in the dessert was very hot and dry with sandy winds. The officers here were much friendlier than before and we could freely talk with them. There was no need for "Yes, sir!" or "No, sir!" I hoped to make a good friend so we could each have an ally. That night after dinner, a young lieutenant named Iyov approached us, and invited us for a ten-minute night run around the base. We were running in total darkness when he suddenly stopped the group. At that place, a heap of wood had been prepared in advance and he set it on fire.

He asked us to sit near the fire, and in its light, he delivered a friendly speech. He explained to us that we were a group chosen to be properly trained for a month and a half here at the base. In the month of July, we would move to the front line to replace another group from this base who were already there for three months right under Egyptian fire. It sounded very difficult to me, but as a soldier, I couldn't say that it was too much for me. He spoke very warmly, but he was actually giving orders. This lieutenant was aware of our feelings after hearing this and tried to raise our morale. He made us sing in a chorus around the fire and we sang for over an hour. We then ran back to the base and fell asleep in our tents.

The day after, we woke up for breakfast and then met a captain named Menashe Inbar. He had already served his three years of mandatory service and signed up to be a career officer. He had the appropriate traits of an officer and was a strong man. He trained many soldiers for combat. He started his speech by saying that he wanted to start with the good news and finish with the bad news. The good news was that we would receive a two-week vacation to go home which would allow us to celebrate Independence Day. The bad news was that we would pass very difficult trainings and in July, we would join the real war at the central zone of the canal.

Our platoon was part of a tank battalion and was designated platoon number 10. This battalion stayed ten kilometers away from the canal, while we in the infantry had to protect the bunkers there. The captain continued to explain to us that the vacation was given for the purpose that we would return fresh and strong for the upcoming training before the war. He wished us a pleasant vacation. We listened carefully to the speech of Captain Inbar and felt that he was right. On one hand, everyone was happy about the vacation, but on the other hand, we felt sad to know where we would be coming back to; a place where there were casualties daily.

At that time, two weeks of vacation seemed like a very long period for me, and I imagined that maybe after we returned there would be peace between Israel and Egypt, and the Army would not need us there. It was on a Thursday that we left for vacation and as we saw the buses approach to pick us up. The morale of the soldiers skyrocketed. Every soldier needed a pass to get on the bus and towards evening I arrived home.

As usual, I was welcomed warmly and told my parents that I would stay with them for two weeks. At that point, I did not want to tell my parents where I would have to go after the vacation. I didn't want to think about it myself. I wanted to enjoy this vacation as much as possible. I rested and hung out with my friends and family in Tel Aviv. I visited my last place of work, the Relif factory, and was warmly welcomed by my former boss, Pozner. To my surprise, he offered me to work a few hours a day during the vacation from the army in order to make a bit of money. I gently declined his offer and told him that I intended to use each free minute to hang out and rest. It was the days before the twenty-second annual Independence Day of Israel and the whole country was in the stages of preparation for this event. Every public place displayed the Israeli flag and was decorated with lights and colors.

My friends invited me for a sequence of night-time performances in Tel Aviv. During the days, I went with my family to the national park in Ramat Gan where my brothers-in-law Tzvi and Asher prepared a barbecue. The park was overcrowded. I was having such a good time that I simply forgot that I was serving in the army. The time passed very quickly. I was ready to wear my army uniform again and return to my base and from there to the actual war. I encouraged myself not to worry and to be happy and to trust in G-d's help.

On the day of my return, I left home dressed in my military uniform with my weapon on my arm. I went to the pickup point for the soldiers and boarded the bus for the long journey to the center of Sinai. I felt calm and relaxed after this enjoyable vacation. I hoped to make new friends with good personalities upon my return. The bus traveled very quickly and passed many towns and cities until it reached the final one, the city of Ashkelon before entering the Sinai desert. The bus driver stopped here and announced a half hour break for lunch. Since the ride was very long, the army equipped the bus with two drivers who switched after the lunch break.

Before evening, we arrived at the base in Birt Madeh. The officers had expected our arrival and lunch was ready on the tables. Right after the meal, we were summoned to a meeting with the platoon commander. He delivered a short speech explaining to us that difficult training would start tomorrow morning. He then took us for a night run for an hour. After the two week vacation, we were a bit out of shape and returned to our tents exhausted.

At the time, the national Israeli soccer team attended the world finals in Mexico and because of the time differences, the games took place late at night in Israel. Most of the soldiers were soccer fans. They listened to the radio for direct broadcast of the games until late at night. The noise in the tent from the radio and of the excited soldiers together, with the cigarette smoke prevented the soldiers from sleeping well. Naturally, they had difficulty waking up the next morning for the advanced training.

All the training was done in open spaces, mainly with armored vehicles. The area of the desert was made up of small hills and loose sand. It was the month of June and the summer there was very hot and dry and therefore, we started the training in the early morning and stopped it around noontime. We then rested until the late hours of the afternoon to escape the heat when it was at its peak. We ate meat for lunch and were forced to drink a lot of water. Besides this, we received hot soup despite the hot and dry desert weather.

Generally speaking, the food was tasty and we had a large appetite. We attended sport competitions and night runs with the platoon commander or another low-rank officer each evening.

The training was difficult and from time to time. The battalion commander and the commander of the brigade made random visits to check on us. This training was done with live ammunition and if the training was unsatisfactory, the officers gave orders to repeat it for another day. If the training passed successfully, the officers would gather us together and praise us. On one of these days, the commander of the brigade was present for the whole day of training. At the end of the day, he delivered a speech. He said that in his view, we were well prepared and ready to go to war against the enemy. The other part of his lecture was about what we were going to do at the front line with the Egyptians. He explained the differences between our forces and those of the Egyptians. He enhanced our self-confidence and morale.

Our breaks to go home were frequent and enjoyable. Every other weekend, we left the base on Thursday and returned on Sunday evening to continue our training. Thus, the time elapsed quickly and in one more month we would move to the canal line to replace the soldiers who had been there for three months. During this training, more experienced soldiers would join in after they had already been at the front line for some time.

Right before our deployment, we received bad news from the front line. A sequence of ambushes against the Israeli forces caused many casualties and deaths. There were daily shellings and we knew that we would be there very soon. Some of the soldiers in my group were very scared and we became friendlier with each other. We had deep, personal discussions amongst ourselves. One of the soldiers, named Biton, said to me, "I have a bad feeling that I will not return alive from the canal." Another soldier said, "I am not afraid to die for the homeland but I have pity for my parents." This type of pessimism was against my views, which were always to think and talk positively. This was part of my religious belief since I had faith in God.

Ten days were left for until our departure. This time was totally dedicated for training that simulated the conditions on the front line. A bunker was specially built in the center of Sinai for these exercises. We were dressed in helmets and flak jackets and everyone received a whistle. The soldier who was on guard duty would blow his whistle the moment he heard the sounds of fired shells. We had to immediately run into the bunker and the guard himself would get into a small bunker which was near his guard post. The purpose of this exercise was to keep us alert and cautious. At the end of this exercise, the commander of the bunker, Lieutenant Amiram, would give us a progress report. He told us that according to what he had seen, if the Egyptians would have invaded our bunker with 150 soldiers, the result would be 100 dead and another 50 would retreat to Egypt. He was very satisfied with our exercises. Each bunker at the front line accommodated ten to fifteen soldiers. In addition, was the commander of the bunker, a sergeant, a medic, and a paramedic.

On the last weekend before our deployment to the front line, we were supposed to go home for vacation. Unfortunately, all leave was canceled at the last minute. During the last two weeks, there were too many casualties on the front line and the commanders feared that if they allowed vacations, some of the soldiers would make excuses to not return. Because of this, we rested at the base on Friday and Saturday. The deployment was planned for Sunday and we were already physically and mentally prepared for it. Each of us knew our duty. We received an order to pack our personal equipment and be ready to move.

On Sunday morning, a truck arrived at the base to take us to the canal. The commander who was in charge of this move, Captain Inbar, drove his own jeep in front of the truck. After a few hours' drive, we arrived at a military camp called Tasa. This camp was ten kilometers away from the bunkers at the front line. We received an order to stay there overnight, and in the early morning before sunshine, to move to our designated bunker. We ate MRE's (meals ready to eat at the field) and we spent the time playing silly, childish games.

At midnight, we prepared ourselves to rest and placed sleeping bags on the sand. Suddenly, Captain Inbar arrived and ordered us to get up immediately since we were moving to the canal that night. The zone was relatively quiet that night and he wanted to take advantage of this for our deployment. We were divided into three groups with a commander for each. Every four to six soldiers boarded an armored vehicle which would take us to our destinations. The journey would be in absolute darkness, which was a very difficult task for the driver. Our designated bunker was located opposite the Egyptian city of Ismaeliya at the center of the canal line. There were a group of other bunkers; one to the north of us and the other to the south. The first armored vehicle went out quietly. It took forty minutes until it safely reached the bunker. Then the second one moved and when it reached the bunker, the last vehicle would go on its way.

I was assigned to the third vehicle. Before boarding, my commanding officer, Amiram, gave his last orders, "Comrades, from this moment on you must wear flak jackets and helmets. Our weapons must be ready the whole way until we reach the bunker. I am asking you not to break down despite the difficulties and not to complain. Anyone who quits is giving an advantage to the enemy. We have to be strong and fight until the end."

The armored vehicle I was in started to move while I was sitting near the communications radio. From time to time, it announced the conditions at the bunker. We drove in absolute darkness since it was forbidden to shine any light. The armored vehicle advanced along the shiny white line in the middle of the road. We lost our way a few times since the white line was partially destroyed by the frequent shellings from the Egyptians. Our commander helped the driver find his way back onto the route. When each armored vehicle arrived at the bunker, it dropped off the soldiers and immediately left to a tank hangar which was located seven kilometers away from the bunker.

When the second armored vehicle came close to the bunker it made a loud noise and the Egyptians, who were on the other side of the canal, immediately responded with shelling. Fortunately, their aim was not accurate and the vehicle reached the bunker safely. However, this incident delayed the third armored vehicle from starting its journey. We reached the bunker at four o' clock in the morning. The bunker sergeant was waiting for us outside with a small flashlight that he used to direct the driver to the entrance of the bunker. He then instructed us to leave all of our equipment on the vehicle and to run out with only our weapons. It was dark inside the bunker and I repeatedly hit my head on the walls as I advanced further. Finally, I found a mattress, put down my helmet and weapon, and rested. I was so exhausted that I fell asleep still wearing my uniform and boots. I knew that if nothing special would happen, I would be able to sleep at least until sunrise when I would be ready for new orders.

The next morning, Yoram the bunker sergeant woke me up asking, "Is it that you Michael?" It was so dark inside, he couldn't recognize who I was. He told me to get up and stand on duty in the guard post. The guard post was in a trench between bunkers. He ordered me to wear a flak jacket and helmet and to have a loaded weapon ready. With a pair of binoculars I had to observe what was going on in the Egyptian side. Yoram warned me that it was strictly forbidden to expose my head, because if I did, an Egyptian sniper would be ready to shoot.

I took a prayer book with me for the morning prayer and entered my post. I looked through my post's barrier and saw the Suez Canal with its quiet, blue water at a distance about fifty meters from me. On the other side of the canal, I saw the city of Ismaeliya with its buildings and gardens. Its three skyscrapers were demolished by the bombardment of the Israeli air force. Egyptian soldiers in uniform wandered from place to place with no protection whatsoever. They had no flak jackets, helmets, or even weapons. I saw cannons and mortars being moved from place to place. The distance between us was no more than 150 meters. Every ten minutes, I closely observed their activity and made a report of what was going on in the Egyptian side.

Every spare minute I had was dedicated to my morning prayer and reciting the psalms of King David for the success and well-being of our soldiers. My morning shift lasted for three hours. For the night shift, I was in another guard post together with another soldier, my old friend Shemesh. During the night, we had to guard our side to prevent Egyptian raids. Many times, the Egyptians shelled our bunkers so terribly that they shook. In daytime, the Israeli air force would immediately retaliate at the source of the shelling in order to stop them. At night, this was done by the Israeli tanks. The tanks would approach from behind and bomb the other side thoroughly until there was no activity. This was the constant routine while we were serving in the bunkers.

We had to be very cautious with our safety and each soldier was responsible for his own life. Anyone who disregarded the rules would lose his life like a friend of ours whom we nicknamed 'Champion'. He was a volunteer from France who thought nothing bad could happen to him. One day, he left the bunker to go to the restroom. The guards who were on duty at their posts heard the firing of cannons from the Egyptian side. It usually took fifteen seconds until the projectile arrived at our position. The guards whistled as a warning for everyone to immediately take cover in the bunker. 'Champion' disregarded this whistle and unfortunately, he was killed at the spot. This was a terrible day for our group. All of his friends in the bunker were stunned and pale. We had been there for just one week and already had a death on our side. I didn't leave the bunker until the corpse was removed. I did not want to see this tragedy.

The kitchen was at a far distance from our bunker and served a few bunkers in the central zone. Every day, two soldiers from our group would take turns going to the kitchen to bring food for the rest of us. A delivery truck brought fresh food to the kitchen and two soldiers named Sasson and Nachesi, who were professional cooks, prepared us very delicious meals. The real problem was to bring those meals into the bunker. When the meal was ready we received a phone call to come and pick it up. Only in the case of no shellings, two soldiers of our bunker went very quickly to pick up the food. We would then sit and eat and afterwards we would rest until it was time for guard duty. Sometimes we would be bombarded while we were carrying the food trays. We could not run through the trenches with these trays so we had to toss them aside and return empty-handed. When this happened, we either had to eat our MRE's or nothing at all.

Sometimes there were consecutive days of constant bombardments. The delivery truck was not able arrive under those conditions and we had nothing to eat. We endured many difficulties and frightening moments. Some of my friends wept out of fear. We were entitled to a weekend vacation once every three weeks. Once a week, a temporary telephone line was installed in the bunker and the soldiers could have contact with their families. My parents did not have a telephone at home at that time. I mailed postcards to my family members every two to three weeks. My cousin Miriam constantly sent me encouraging letters and I always responded to them. In my letters, I never wrote where I was located or how difficult things were for me. I did not want my parents to be worried. Nevertheless, they knew a lot about my service from friends of mine. I always said to them that everything was okay and that I had to fulfill my duties in the army. A tragedy could also happen when you serve close to your home.

During the night, besides guard duty, we would fill sand bags in order to improve our posts. Those posts were damaged from the constant shelling. Once a week, a civilian bulldozer would appear to raise an artificial hill so the Egyptians could not see us. We worked physically hard and were on alert all the time. At night, we feared we would be raided by the enemy. They tried to raid us many times but failed. Our defensive strategy against these raids was successful. The guards would have spotted the enemy as they swam across the Suez Canal. When they emerged from the water, we would let them get close to the bunker and then fire and throw grenades at them. The survivors of our assault would retreat. They could never accomplish even a single raid successfully. Nevertheless, these frequent incidents caused us to be nervous at night. We never knew when they would come or if we would always prevail over them.

At the end of July, the American administration tried to mediate between the two opposing sides. U.S. Secretary of State Rogers came to the Middle East with a plan to reach a ceasefire for three months in which the sides would sit together and talk. Only through peaceful negotiations could Egypt retrieve the land they lost in the Six Day War. By using force, they had no chance to get anything back from Israel. As a matter of fact, the Egyptians were using force for a long time and achieved nothing of substance. They bombarded us day after day and night after night. On Saturdays, they especially intensified their bombardments.

As long as the border was quiet, we exited the bunker cautiously to reinforce our posts with sandbags. We also helped a special unit of civil engineers who came frequently to structurally improve our posts. This task was quite difficult physically since we filled the sandbags one after the other while wearing our helmets, flak jackets, and weapons. Amiram, the bunker commander himself would join us in this task and said to us, "Each bag protects our lives as it lessens the probability of being wounded."

One night, Commander Amiram was outside of the command bunker while the Egyptians delivered heavy bombardment in our direction. He had disappeared and no one knew where he was. At such a crucial time, we were in need of his instructions and orders but his voice was not heard through our communication system. Usually, he used a communication device at the command bunker to check the safety of each soldier at the guard post. However, this time we had to look for him since we did not know what happened to him. Unfortunately, we could not get out of the bunker as the Egyptian bombardment had no end. Suddenly, there was a two minute period of silence and then I saw and heard someone running towards the entrance of our bunker. I demanded this person to identify himself and I immediately heard the response, "It is me, Amiram!" At that moment, I felt great relief and he immediately came in and sat across from me on the bed. He was exhausted from his run and smiled to me. He asked, "Do you know where I have been?" I told him that I was curious to know since we were all worried about him.

He then told me that for half an hour he was lying on the ground in one of the trenches and could not raise his head. The Egyptian bombardment was very heavy and any movement would have risked his life. He then asked me for a cigarette which I gave and lit for him. I then advised him to call the command bunker since they were looking for him. He immediately called the command bunker as well as all the guard posts and all the other bunkers in the central zone. He asked them if everything was alright with them. He was chain smoking one cigarette after another very nervously, with the communication device in his other hand. He reported to everyone that he was okay and that he was located in our bunker, number three.

On this occasion of being this close with my commander, he informed me that if the Egyptians intensify their bombardments, it's a sign that they were planning a raid. The Egyptians knew that all of us were hiding in the bunkers but we were smart enough to frequently check what was going on in the water of the Suez Canal. The guards at their posts were on alert at all times. Even as I lied on a mattress in my bunker, I would peek outside at the canal to make sure no Egyptian soldiers were swimming towards us. All of us were very cautious. Amiram finished his last cigarette and ran to the command bunker. From there, he reported in detail to the tank unit what was going on in our location and they immediately retaliated for over half an hour with heavy fire. This determined response calmed down the Egyptian forces that were constantly active for over forty minutes.

It was my turn to stand at the guard post with my friend Shemesh. We were well-equipped and carefully observed the water of the Suez Canal. During those moments, we listened to the bombings performed by our tanks and saw flames on the Egyptian side. My friend Shemesh was overjoyed and laughed loudly. He said, "Now we are teaching them a lesson."

I was standing behind my friend when we suddenly heard the whistle of an Egyptian shell. We immediately jumped towards our *shafania* which was part of the guard post. Shemesh was the first to get in. I followed him and before my whole body was in the safety of the bunker, the shell had already exploded. I suffered from shrapnel wounds in my left leg and the blast shocked my body. I was overwhelmed with fear and I trembled. Shemesh asked me if I was alright and I answered, “Yes and no.” I don’t know how long we stayed in the *shafania* but when the zone had become quiet I asked Shemesh to help me get up. He told me to wait until he checked if there were any raids coming in our direction. He also reminded me that in a few minutes new guards would come to replace us.

We left our guard post on time and I walked with a limp while clutching the side of the trench. Sergeant Maoz met us on our way and asked what was wrong with me. I told him what happened and he accompanied me to the command bunker where a physician and a nurse were always on duty. When we approached the entrance to the bunker, the sergeant shouted, “Doctor, you have a patient on the way.” I was placed on a stretcher and my pants were removed. At that moment, I didn’t feel any pain, but only warmth in my wounded leg. My undergarments were covered in blood. When the bunker commander Amiram, who was on location, saw my bloody undergarments he wanted to have a chopper evacuate me to a hospital. The physician examined the wound very carefully and then determined that it was not life-threatening. He then said to Amiram that he can take care of the wound and there was no need to put a chopper at risk. The doctor removed the shrapnel from the upper part of my leg and he treated the wound.

The commander approached and asked me if I was alright. I answered, “Yes, I am alright.” He then asked if I was able to stand on guard duty when it was my turn. “Yes”, I answered. “I do not think that my injury is serious since I am not in pain.” I stayed in the command bunker for another half hour and then stood on my feet and went to my bunker which was nearby. I tried to sit down but I could not because the pain had gotten worse. I lied on the bed and the pain stopped. Anyhow, a while later the pain suddenly returned, even though I was resting. My friends wanted to notify the physician and commander but I discouraged them from doing so. I told them that everything would be alright and the pain would pass. I did not want another soldier to stand guard in my place.

That night, I took my turn at the guard post as usual and finished it on time. I fell asleep from exhaustion and woke up in the morning in a state of panic. I went to the physician to have my bandage changed. He asked me when it was my turn to go on leave. I answered him that my turn would come in the next few days. He gave me a referral to be examined in a hospital. I said that I hoped I would not need to go to a hospital. He smiled and said that if another soldier and another physician were involved I would have been evacuated to a hospital as soon as I was injured. He concluded, "I really appreciate your behavior."

That same night, the whole canal line was under heavy bombardment by the Egyptians. In two bunkers that were adjacent to ours, two soldiers were injured. One of them was a platoon commander who was injured in the foot and was immediately evacuated to a hospital. The other soldier was a staff sergeant who was injured in his eye and was also evacuated. Our bunker commander Amiram gathered us to the command bunker and delivered a speech to us in which he analyzed how those soldiers were injured. The purpose was to teach us a lesson. At the end he said, "Comrades, do not give up and stay in high spirits even though this is not easy at all."

Three soldiers on leave had to return that week, so that three others would be able to go on vacation. I was among the three who were permitted to leave. As I finished one of my guard shifts, I saw an armored vehicle arrive with the three returning soldiers. The bunker sergeant ran to me and told me to immediately put on my dress uniform and be ready to leave. I went into my bunker to change and filled my duffel bag with laundry to take home. I boarded the armored vehicle with my helmet and weapon. The sergeant waved me goodbye and told me to take care of myself and enjoy my vacation. I received a vacation period of a whole week. We left the bunker Thursday evening and after fifteen minutes we reached Camp Tasa. Unfortunately, all military transportation that picks up the soldiers had already left. I needed to hitchhike and knew that it would take me six to eight hours to reach home this way. I stood on the side of the main road to Refidim, which is in the center of the Sinai Dessert. When I reached Refidim, I had to hitchhike again to El Arish in the Gaza Strip.

An Israeli motorist driving a long semi-trailer picked me up and dropped me at El Arish. It was already late at night when I saw a restaurant that served Israeli soldiers. Many of them were sitting there eating dinner and I joined them. When I finished I again stood next to the road to hitchhike to Tel Aviv. After waiting for half an hour, the same semi-trailer arrived. The driver stopped again and asked me if I needed to go to Tel Aviv. When I confirmed that I did, he said he was ready to take me there under the condition that I do not sleep on the way. I immediately agreed and sat near him. He turned on the radio at a high volume and kept talking with me so he would not fall asleep himself. I reached home by two o' clock in the morning.

I knocked at the door and my parents woke up immediately. Dad got out of bed and opened the door for me. Mom said from her bed, "I knew that it was you. Welcome home. All of us are sleeping and we will talk tomorrow." I urged Dad to go back to bed since he had to wake up early for work. I undressed and saw that my undergarments were still stained with blood. I put on my pajamas and immediately fell asleep until ten o' clock in the morning. I woke up from the smell of Mom's cooking for Shabbat. I especially liked the smell of the fried fish and eggs of the cooked *cholent, Palow shabati.* Mom liked to start her cooking in early Friday morning so that everything was ready by noontime.

I rose out of bed and took a shower while Mom was preparing my breakfast. Before I ate, she told me that a few days ago she had a bad dream in which she saw me in a very dangerous place where I had gotten hurt. The day after she had this dream, she went to the butcher in the market of Yahud. She bought a live chicken and took it to the slaughterer as a *kapara.* She demanded from me, "Please tell me the truth of what exactly happened to you." "Nothing special," I said. "You can see that I am healthy and whole and am here for my vacation. I am in the army and it is not easy to serve there but this period in my life will eventually pass." She was not satisfied with this answer and continued to talk about the risk and dangers of my military life.

She said, "We have been very worried about you and I went to

Grandpa to discuss our worries about you. He told me not to worry since nothing bad would happen to you. He always prays for you." I said to her, "Yes Mom, I know that Grandpa is a righteous man and his prayers are not in vain." She asked me, "For how long are staying with us?" I answered that I would be with them for a whole week. "That is not enough at all!" she responded. "However," she continued, "I hope that by the end of your vacation peace would come and all soldiers would be discharged." I said, "Hopefully."

I helped Mom prepare for the Shabbat and then went out to wander the market. Friday is usually considered a short day since it is forbidden to work by sundown in our religion. Since this was summer time in the month of July, the days were longer so I was able to spend more time outdoors. I rested a lot during this vacation and was happy to be surrounded by my brothers who were on vacation themselves from school. They asked me many questions about what was going on in the army. I did my best to satisfy their curiosity by answering all their questions as much as possible. I went to visit my relatives and my free week was passing by very fast.

I still had to be examined at a hospital as ordered by my physician's referral. However, I had no desire to undergo this examination during my vacation. I still had pain in the upper part of my left leg and beneath my stomach and I hoped that it would pass by itself. Even at the end of my vacation, I did not go to the hospital.

CHAPTER 21

My week of vacation had already passed and I had to go back to the Suez Canal. My colleagues were eagerly waiting for my return so they could go on vacation themselves. On the news, there were many discussions about an upcoming ceasefire agreement between Israel and Egypt. U.S. Secretary of State Rogers had mediated the negotiations between both sides. He was highly trusted by the Israelis. All that was needed was the Secretary's offer to be accepted by Egypt.

I bid farewell to my parents and family and took a bus straight to Camp Tasa where my battalion was located. There, I joined some of my other friends who had also returned from their vacation and boarded an armored vehicle to my bunker. As a precaution, all of us had to sit in the vehicle wearing our helmets and flak jackets. When we arrived, the commander welcomed us back warmly. He immediately referred me to the physician who checked if I went to a hospital for my required examination and X-rays. I told him that I did not go since I had no time and I did not think it was necessary. The physician got very angry and yelled at me. I told my friends what happened during my meeting with the physician. They said that I was not normal and crazy. They added,

"Anyone else in that situation would impersonate as a cripple and get discharged from the army. The minimum you could have achieved was to serve a non-combat job near your home." I answered, "That is not me. I am not like that. Don't you guys need my help?"

The next few days were quiet on the front line. However, the

Egyptians bombarded us at night. This was a real problem that made us afraid. We had to be on alert at all times and to check the borderline very carefully. When we discovered footsteps in the area between the bunker and the canal, all the guards at the Israeli side were searching for the infiltrators throughout the whole borderline. When they were located, they were shot and killed. Whenever it was quiet, either during the day or night, we were busy filling up sandbags. At that time, I felt pain from my injuries. Therefore, every free minute I had was devoted for rest.

One day, we worked outside as usual, and we suddenly heard cries of agony on the Egyptian side. *"Ya Raees!"* At the same time, the Egyptian radio announced the sudden death of their president, Gamal Abdul Natzer. He passed away in his early fifties due to a heart attack. This president was an army officer who seized power by overthrowing the monarchy about twenty years earlier. He was an Egyptian nationalist and a bitter foe of Israel. Anyhow, we did not celebrate his death since he signed the ceasefire agreement himself right before his passing. An unknown Egyptian officer, Anwar Saadat, replaced Natzer and he strictly adhered to the terms of the agreement. Our quality of life changed dramatically afterwards. This ceasefire agreement was in effect for three months as was stipulated.

In reality, Egypt needed this ceasefire more than we did. They suffered a lot of casualties and property damage. Israel needed this ceasefire to reinforce the borderline. The line of bunkers parallel to the east side of the Suez Canal was called the Bar Lev Line. Lieutenant General Bar Lev was the Chief of Staff of the Israeli Army. It was his idea to build bunkers parallel to the canal line. Now we had to exploit these three months to repair the damages inflicted on our bunkers and guard posts as well as reinforce them. Besides, new bunkers were planned to be built quickly. These bunkers directly opposite the Egyptian forces were called Maozim. Bunkers on a secondary line at a distance of ten to fifteen kilometers from the canal, called Taozim, were already being established while the ceasefire started.

Israeli civilian contractors were responsible for this quick construction with the help of both civilian and military labor. All highways in the Sinai desert were flooded with large trucks carrying building materials as well as cranes and bulldozers. We took part in this tremendous undertaking which exhausted physically. Nevertheless, this was considered a good period for us since there was no fear of bombardments or infiltrations.

The was August 1970 and there were very hot days near the canal. We didn't have to wear helmets and flak jackets but we still had our weapons ready. One more month was left until the end of the ceasefire agreement and my group had finished its share of labor. Afterwards, we did not have much to do besides standing guard. In our free time, we built for ourselves a volleyball court and used to play there for two hours daily. I could not play myself because of my injury. I would be wracked in pain from time to time. My friends decided to appoint me as a referee. I simply sat on a chair posted on a hill with a whistle in my mouth and used it from time to time in favor or against a team.

The army sent many entertainers to perform for the soldiers at the front line. The most popular Israeli singers came to the Suez Canal line. We watched a movie almost every day. At that time, we were able to eat regularly and the quality of the food improved. A new professional cook joined the kitchen as well as a kosher supervisor who was a Hasidic Jew. An atmosphere of peace and tranquility befell on us. Each night, we sat together with our commander Amiram and we would tell stories and jokes. This was an occasion for us to get to know each other better.

Some of my friends proved to be extremely funny, especially one who we all called Napoleon, and never knew his real name. This soldier was very short and always acted strangely. He would always keep his distance from us and would not take part in our games or attend the movies we watched. One day, all of us were sitting in the bunker watching a thrilling movie when the electricity suddenly shut off. The movie projectionist checked all the connections of the projector and did not find any defects. He then left the bunker to check the generator. He discovered to his surprise that Napoleon had taken out the plug of the projector and put the plug of his electric razor in its place. Napoleon was shaving and singing at the same time. The projectionist yelled at Napoleon, "What are you doing?! All of your friends are watching a movie and you are ruining it." Napoleon laughed loudly and said, "Wait a minute or two until I finish shaving." We experienced many funny events like this.

It was unbelievable to realize how peacetime can cause changes to a person. We received mail very often and newspapers daily. Vans that belonged to *shekem* arrived with cigarettes, sweets, and drinks. Because of this extreme change we felt like we were on vacation. I imagined to myself that on the other side of the canal, our Egyptian counterparts were happier than us from the situation. They had suffered much more casualties than us before the ceasefire got into effect. Hence, the change for them was more significant. On the other hand I thought, they did not care so much about human life. All that interested them was retrieving the pride they lost after the Six Day War. The whole time since then, they declared, "What was taken from us by force we have to retrieve by force." The position of the Egyptians was supported by the Soviet Union which was a super power. They constantly supplied the Egyptian army with new weapons and encouraged them to continue fighting with Israel.

The pains from my injury did not totally stop and returned from time to time. Yossi, our paramedic determined that I had to see a specialist for a thorough exam. He suspected that the constant pain was a symptom of something severe inside the body that had to be immediately diagnosed. He told me, "Do not to take this pain for granted. You could never know the cause of it." A specialist physician served in a military base in Refidim which was in the center of Sinai. Since this was a period of peace and I was not missed at the front line, I accepted the paramedic's offer without protest. The bunker's physician gave me a referral for that specialist and the next morning, I traveled to see him.

I arrived at the specialist physician's office in Refidim. He asked several questions regarding my injury like how it happened and where. He then thoroughly checked the location of the wound and said that he had to refer me to Assaf Harofeh State Hospital south of Tel Aviv. He suspected that I had a severe fracture and it was possible that I needed surgery. To be sure his suspicion was correct he needed the hospital to perform X-rays. "What!? Surgery?" I asked incredulously. "Yes, young man." He said. "Why did you wait until now? It has to be treated immediately since it could get worse." The physician wrote down a full report about my injury and enclosed with it my whole medical file to be submitted for the hospital.

The next day, I arrived at Assaf Harofeh Hospital where I was subjected to a series of exams. These procedures lasted for almost the whole day with a lot of waiting. The final conclusion was that I had to undergo surgery. The physician calmed me down and said, "Listen. This is an easy surgery and you do not have to worry. I will set you up an appointment for this surgery as soon as possible. In the mean time, you are free to go and come back next week." I left the hospital and did not know where to go from there. I had the choice of going back home, which was not far, or returning to the base. Naturally, it was preferable for me to go home but I did not know how to tell my parents about my health problems. After thinking for a while, I decided to go home. I told my parents that I had received another week of vacation and I wanted to use that time for rest since I was exhausted from hard labor. I did not mention the truth about my arrival.

During that whole week, I thought about my upcoming surgery but acted normally around my family like nothing was wrong. At the end of the week, my friends were scheduled to leave the bunkers and move to the military base in Refidim. This was a normal base with small buildings and rooms. Next to it was a huge air force base and service center. Plenty of soldiers served at that location in the center of Sinai. At that same time, I had to appear before a physician at the hospital. This was not pleasant for me but I had no choice.

This was my life and I had to take care of myself. To be frank, I did not know what a fracture was, but I knew that it had to be treated. I returned to the physician at the hospital and he gave me a letter to deliver to the operating room. The chief nurse there asked me who would be accompanying me. I answered, "No one except G-d." She asked in astonishment, "Your family doesn't know about your procedure? Do you have any family?" I answered no. She said to me, "Tomorrow or the day after tomorrow you will have this surgery here. However, it may be possible for you to have it this evening. It's now peace time and we don't have any casualties besides accidents. Are you ready for the operation?" I answered yes. I was eager to pass the ordeal already. She assigned me to a room where I had to dress in a patient's gown.

I lay in bed for two days at the hospital. The night before the surgery, a nurse approached me and said, "Michael, come with me so I can prepare you for the surgery." The nurse cleaned me up and shaved off all the hair on the location of my injury. He then warned me not to eat or drink until tomorrow morning. The next morning at nine o' clock, another nurse appeared and placed me on a rolling stretcher. He took me to the operating room while I was still conscious and aware of my surroundings. In this room I saw two physicians and two nurses before I was administered an anesthetic. The last words I heard before I fell asleep were one physician telling the other that it was okay to begin.

In the middle of the night, I found myself in the recovery room with a few more soldiers who had their mothers with them. Apparently, while I was regaining consciousness in the recovery room, I moaned very loudly as if I needed help. One of the mothers thought she was helping me by sprinkling water on my face and lips. When I opened my eyes, I saw this woman standing in front of me and she said, "Soldier, you are okay. Who is coming to help you?" I did not respond. I just thanked her for her kindness.

As the effects of the anesthetic wore off, I started to feel severe pain. A nurse was called to my bed and she brought with her a drink and painkiller. She instructed me not to get off the bed or even not to move. She added that after such a surgery I could not eat solid food. I told her that I understood everything. From then on, whenever I felt pain, I asked for a drink. It would soothe the pain especially at night when I could not sleep. The hospital staff attended my needs very nicely until I recovered.

Three days after I woke in the recovery room, a telephone was brought into that room so patients can call home. The first person I called was my brother Amnon. He answered "Hi Michael!" with enthusiasm. Then I started to talk and he immediately interrupted saying that I did not sound good. "Yes Amnon," I responded. "I am now in Assaf Harofeh Hospital." He asked if I was alright and which unit I was staying. I calmed him down and told him that I have just had surgery and the worst was behind me. I was feeling well and everything was alright. He then asked me if the family could come and visit me. "Yes," I answered, "As long as it is not difficult for you."

Amnon asked me if I had told Mom or anyone else at home about my ordeal. I told him that no one knew about it. I assured him again that everything was alright with me and there was nothing to worry about. A few hours later, I received a big surprise. Mom and Amnon had arrived to visit me. My brother was carrying a big bowl with some of Mom's delicious food. The bowl contained pieces of meat, fruits and vegetables, dried fruit, as well as sweets and chocolates. Mom asked what had happened and why did I not tell my parents anything when I recently visited them. I answered that I felt it was better not to tell them so that they wouldn't worry. Mom urged me to eat immediately and I told her I was not allowed to eat solid food. I then asked my brother to ask the nurse if I was able to eat yet.

Amnon instead went to talk with the physician in charge and received an explanation of my condition. The doctor told him that I could now start eating but only very soft food. One hour after my family came to visit, the chief nurse entered the room and said that visitation time was over. Mom asked her when I was going to be discharged. The nurse checked my medical file and said that if everything was alright I would be home by Shabbat.

The whole unit in the hospital already knew that I did not get many visitors. A personal nurse was assigned to help me go to the restroom. Three days before discharge, I was already feeling very well. They removed all of the tubes from my arm and I was allowed to eat normally. The chief nurse told me that it seemed I would be discharged on Friday.

I used this time to wander alone in the hospital garden.

On Friday morning, I was approached again by the chief nurse. She told me that I was able to go home on that day. I carefully dressed up in my military clothes and went to the unit's secretary to receive a discharge note together with my medical file. Before leaving the hospital, I had to go to the military physician's office for instructions. He informed that I have to be cautious for six months and referred me to the *Bet Marpeh* for three weeks. The house was located on Mount Carmel overlooking the beautiful harbor of Haifa.

I returned home by hitchhiking. My family treated me as a sick patient who had to be spoiled. I felt sticky and wanted to take a shower, but I was covered in bandages. These bandages were wet from blood and puss. I also wanted to replace my uniform with convenient civilian clothing but did not know how to manage all of this. I called Amnon and asked him what to do. Amnon immediately soothed my worries. He told me that he would go to the pharmacy to buy new bandages and bring them to me. Until his arrival, he recommended me to take a shower with warm water and thoroughly clean the location of my wound. I accepted his suggestion and when he arrived at my parents' home, I was ready to have my bandage replaced. I felt very weak but happy to be with my family again. I rested at home for the entire Shabbat and had to be at the *Bet Marpeh* by Sunday.

I left home on Sunday morning and again hitchhiked my way to the northern city of Haifa. When I arrived at the resting house I met other soldiers there who had recovered from injuries. The receptionist assigned me a room on the second floor that had a beautiful view. The building was located on a high mountain surrounded by lawns and trees. At the bottom of the mountain, I could see large parts of the city of Haifa and the Mediterranean on the horizon. This place gave me a wonderful feeling after I was exhausted for nine months from my military service. I had never before been to a hotel or rest house for vacation.

Every week, I received a checkup exam from a physician and treatment from a nurse. We were fed free, delicious food for as much as we wanted. Entertainers were invited to perform for us at the *Bet Marpeh* and we also visited the city theatre to see shows. One night, we even attended the Israeli premiere of the movie "Love Story." The mayor of Haifa was present at this premiere. For three weeks, we injured soldiers enjoyed our time immensely. The resting house staff did all that was possible to spoil us the whole time.

Towards the end of the third week at the house, I was thoroughly checked by the physician of the compound. He said to me, "Michael, you have to stay here for another two weeks until your wound heals completely. The area surrounding of the wound is still infected and swollen and you cannot return to your base in Sinai in such a condition." I happily answered that I was willing to stay longer and I was not against his offer. I didn't suffer from extraordinary pains and I just rested as much as possible.

From time to time, we went out to wander the surrounding area. One Saturday we were allowed to go home for the weekend. Almost two months had elapsed since I had entered the *Bet Marpeh* and was allowed to leave. I received a letter from the physician that exempted me from heavy lifting and hard labor. He also gave me a medical report and instructed me to return to the military base in central Sinai and submit this report to the administrative officer.

I returned to Refidim and after the letter and medical report were examined, my superiors decided to switch me to another battalion. That battalion also belonged to the armored forces and it was number nine. I was instructed to report to an officer named Gabi and was told that he knew what to do with me. I slept overnight at Refidim and in the early morning, I traveled to my new battalion by hitchhiking. The battalion was at a place called Baloza and the military camp was named Hatzav.

I entered the camp which was located between two hills and surrounded by palm trees. The high command of battalion nine was located in this camp. The combat soldiers of this battalion were stationed in the Taozim, the secondary line of bunkers from the Suez Canal. I entered the tent of the administrative officer who welcomed me very nicely.

The name of this officer was Gabi Klod. He read the letter carefully and said to me, "For the time being you cannot be a combat soldier but since I need help in my office, you can assist me here." This office was comprised of a sergeant in charge of religious affairs and another two female soldiers named Yochi and Miri. I asked him, "What would be my duties here?" He answered, "You will be an NCO (non-commissioned officer) in charge of flights and mail." I told him that I had no experience in these matters. He said, "Don't worry. You will learn it over time. It is not as difficult to learn as the Torah of Sinai." I asked him, "But aren't we already in Sinai?" Gabi got serious and said, "I am not joking with you. You have to understand that this task is not difficult. Just do as you are instructed then you will learn everything within a short time." I gave my consent for the job.

I looked over everything the office, which was a large tent, and then I went out to check the second tent which served as a dormitory. A few empty beds were left and I picked one of them for myself. Afterwards I went out to make acquaintances with the whole camp when I suddenly someone yell, "Hey soldier! Come here!" I turned around and saw a red-haired man with a pot belly. I immediately understood from his appearance that he was the First Sergeant in charge of the discipline of the camp.

I approached him and he asked me who I was. I said that I was the new NCO in charge of flights. "Oh really?" he asked. "Stand at attention when you talk to me!" he sternly ordered me. "Yes sir!" I said. He then continued to ask, "In charge of flights? If I need a flight I have to get your approval?" "Yes sir!" I answered. "Okay," he said in a calmer tone. "I want to be in the first flight that leaves this camp on Thursday." I approved his request. He then smiled at me. At the time when we were talking, I knew nothing about flights from there, and I didn't even have the schedule of the flights. All of this was a complicated matter that only the administrative officer could take care of.

All of the soldiers and officers wanted to be on the first flight and there were many quarrels in this regard. In the end, the battalion officer,

Giora Chaike, decided to take care of it. He would decide the order of who would be on each flight. Since then, all the quarreling stopped. On my first Monday on the base, I received the schedule of the flights leaving the airport of Baloza. There were one to two military flights to Tel Aviv each hour. We were very busy working in the office. It did not take a long time until I knew all the necessary tasks.

During my free time in the office, I used to type letters to my cousin Miriam until one day, the administrative officer screamed at me, "Michael, you are going to ruin the typewriter! The female soldiers do not have anything else to type with!" Officer Gabi was a calm, quiet, and very shy person. The female soldiers at the office always took advantage of him and would get off from work. We stayed at that base for about three months and then moved to a location further away from the front line.

This new military base was at Birt Madeh in central Sinai. There were small buildings and tents where the soldiers would reside and a huge dining hall. The cooks in the kitchen were of Moroccan origin and they prepared Oriental food for us. The work here got more and more intensive. I was promoted and had become the administrative officer's assistant. The office staff grew larger with the addition of a new soldier named Yehoshua. I strengthened the discipline in the office and the staff there had to arrive at work at exactly nine a.m. They were no longer allowed to leave the office without a discharge permit. Officer Gabi liked this.

One evening, I was exhausted from work. I went to rest in my tent. There were three beds and I had trouble preparing mine because of the darkness. I could hardly see anything since the only available light came from outside. I suddenly heard a soldier speak from the bed next to me. "Hey! Who is this? What's your name?" I answered, "I am Michael. Who are you?" The soldier said that his name was Malachi Chizkia and he was pleased to meet me. "Where are you from?" he asked me. I answered, "Let's sleep now and we'll talk tomorrow." Malachi responded, "I have already slept for the entire day and I cannot fall asleep now. Anyhow, what is your origin?" I told him I was from Afghanistan. "What? This is the first time

I have ever met someone from Afghanistan. Is there a community of Jews there?" I told him that I was living proof. He asked, "How many Jews live there now?" I answered, "Sorry. I am tired and I want to sleep. We will talk tomorrow." Malachi continued to mutter to himself, "Afghanistan! How interesting. I've never met this guy before, but from the way he talks, he must be interesting."

When I woke up the next morning, I thought Malachi was already outside. However, I then noticed that he had just covered his whole body with his blanket. He liked to sleep in the morning and be awake at night. Malachi was a handsome and intelligent youth and it was pleasant to talk with him. He had good connections with the battalion officer and other officers with the exception of Feldman, the First Sergeant in charge of discipline. Feldman was always looking for him and Malachi always knew how to avoid him. I remember well that he had a special language of his own that no one could understand. You would hear him speak Hebrew but you would not be able to understand him. One day, Malachi, Giora our battalion commander, and I were standing together and talking. Giora liked to hear Malachi's special language. Suddenly, we heard from a far distance the First Sergeant in charge of discipline yelling, "Malachi! Where are you? I have been looking for you for awhile. You are in trouble now!" Malachi did not get scared and responded in his own language which could not be understood. Giora laughed and the First Sergeant got confused and left.

Malachi served with us in the office, since at the front line, a person like him would be out of place. He was with us for a few months and I never knew what he did exactly. I think he was the NCO in charge of education for a while. The administrative officer was very satisfied from the way we worked in the office. He had just one month left to serve and did not want to continue his service further. A new deputy battalion officer arrived at our base called Immanuel Sekal. He was a handsome man but very tough. He started to make our lives difficult.

Six months had elapsed since my surgery and we again moved back to the Hatzav military camp at Baloza, the place between the hills with the palm trees and good weather. I was totally devoted to work and tried my best to be as efficient as possible. I learned almost every task at the office and I could handle every matter. Again, I had to cope with the demands of scheduling people for flights. The First Sergeant always demanded to be on the first flight. Whenever the quarreling became extreme, I involved the battalion commander or his deputy for a resolution.

Many entertainers, military bands, and lecturers passed through our office during that period. My duty was to assign them to the Maozim or Taozim to perform. One military band that used to come to perform often had an act where they made fun of the First Sergeant. One day, their bus arrived and stopped near our office. I then heard the First Sergeant screaming loudly outside, "Michael! Get over here fast!" I went out and asked him, "What happened, First Sergeant?" He said to me, "Go and immediately bring the manager of this band to me." I again asked what happened and he replied, "It is none of your business. Just call him." I went to the band's manager and told him that First Sergeant Feldman wanted to see him urgently, that second. The manager was a little scared and wanted to know what this was about. I told him that I did not know myself and I asked him to accompany me to the First Sergeant's office.

Feldman was angry and immediately shouted at him, "It has come to my attention that in one of your acts, you make fun of First Sergeants. Don't you dare do this again!" The manager tried to protest and explain himself, but the First Sergeant would not even let him open his mouth. When the band's manager left the room, I remained alone with the First Sergeant. He said to me, "They dare to make fun of us. Shame on them! They do not have a drop of respect or education. For G-d's sake, what kind of humor is that? Go and attend their performance to check on them." I tried to explain to him that this was their program, and they couldn't make it shorter or longer, The staff sergeant said, "I don't want to listen to this! Have you heard me?" I responded, "Yes sir!" and left the room.

In the evening, I went to attend their performance for the soldiers at the Taozim. I could not believe what I saw. They did not change anything from their last program, but they added a whole act that made fun of First Sergeant Feldman. The soldiers at the base had plenty of jokes and interesting stories about First Sergeant Feldman. One day, a soldier wanted to go home for three days to attend a celebration with his family. He needed a leave permit signed by the First Sergeant. The soldier approached Feldman and told him that he needed the vacation from the upcoming Sunday and he would return on Wednesday. The First Sergeant did not know how to spell Wednesday. He called the gatekeeper of the base and told him his problem. The guard answered, "I don't know how to spell Wednesday either." Finally, Sergeant Feldman solved this 'problem' by telling the soldier to return on Thursday. We simply gathered together all the funny stories about Feldman. During one of our parties on the base, some of the soldiers told them over and everyone burst out in laughter. I would summarize my service in the base at Baloza as one big party.

CHAPTER 22

One day, the administrative officer received a letter from the high command of the armored forces. The letter was a request for him to stay in the army of another few months since a proper replacement was yet to be found. I also received a letter from the army health authority notifying me that I am already fit to be a combat soldier again since ten moths had elapsed since my surgery. This letter meant that very soon I would be sent back to the front line at the Suez Canal.

Israel and Egypt had extended the ceasefire agreement for a few more months. Thus, the two nations were not in a stage of war. Nevertheless, the Egyptians broke the agreement from time to time by staging raids on our side. The Israeli army was well prepared and there were no chance for the Egyptian forces to surprise us. This was a relatively quiet period, but the Israeli commanders intensified the training of their troops. The commanders had high confidence and morale since they were still under the euphoria of the Six Day War victory.

The deputy commander, Emmanuelle Seckel, summoned me to his office and said, "Michael, you are now fully recovered and you are going back to your original unit." I responded, "What about my duties in the office? Who would perform them like I would?" The deputy commander calmed me down and promised to talk with the administrative officer and to look for a proper replacement. He then left the room.

Gabi, the administrative officer, did not want to accept the decision of the deputy commander. He told me that only the battalion physician could determine my current health status and make such a decision. He said to me, "This physician is a good friend of mine and I always arrange flights for him whenever he wants. Don't worry. I have a meeting with him this week and I will insist that you stay in the office." Now my personal case turned to be an argument between the administrative officer and the deputy commander. The deputy commander said, "If Michael needs a physician's approval to be removed from the office, he should be examined by the brigade physician." He also added that it was unacceptable that a soldier who fits the requirements for combat would serve in an office and there is no argument about this. Naturally, the deputy commander won this dispute.

The next morning, I left the base with my packed duffel bag to Refidim. I went to the office of the brigade and the sergeant there said to me, "You are going to Camp Mitleh where you will meet the administrative officer." I asked what was there, and he answered, "You will get all your explanations there." At that moment, I was filled with despair and felt that I was being thrown from place to place. I left his office and wandered around the giant base. I thought to myself that there was no justice in the military. During my ten months in the administrative office, I did the unbelievable. I fulfilled each task that was assigned to me in the best possible way. Why was I getting thrown to this hole in the desert called Camp Mitleh?

While I was wandering and thinking, I reached the brigade's dining hall and could hear a party going on inside. I did not know what this party was about and this didn't interest me. Finally I decided to enter and attend the party since it would calm me down. I took my place among the soldiers and when I finished eating, I heard someone calling my last name. I stood up and looked around to see the First Sergeant in charge of discipline at this base. His name was Yeshayahu. He was a close friend of my brother Amnon and lived in Yahud. He asked me if I was really Amnon's brother and what I was doing on the base. I answered him that I had just arrived and I did not know exactly what my duty was. I also told him that I felt that people were playing games with me. Yeshayahu went to the brigade's office and enquired about my status. He was informed that I had to move to a new base in Mitleh and he decided to take advantage of my transitional period.

Yeshayahu came to me and asked if I wanted a vacation for a few days since I looked desperate in his perspective. "Sure!" I answered. "I wouldn't object to it." He immediately called the sergeant of the office and instructed him to give me a leave permit for tomorrow morning. Yeshayahu then told me that he was going to give me a vacation for one week and asked me to pass over his best regards to Amnon.

That night, I was very happy and totally calm after the ordeal of the last few days. I said to myself, "Michael, don't worry. G-d is with you and is helping you out." Even though I was tired, I went out in the evening to see a performance at the base by the famous Greek singer Aris Sun. After the show, I was exhausted and went to sleep in one of the vacant tents.

The next morning was *Rosh Chodesh*. For this special occasion, I decided to pray *shacharit* in the brigade synagogue. To my surprise, I found many soldiers there who practiced the Jewish religion strictly. I joined them in prayer. I then went to the administrative office to get my leave permit. The sergeant welcomed me very warmly and asked me how long I would like to have the vacation. I did not know what to answer so he gave me an open pass.

I hung out in Tel Aviv and had an amazing time for two weeks. I then decided that it was not nice to take advantage of anyone who does me such a favor. I decided to return to the brigade's base in Refidim. As I got off the bus there, I saw to my surprise my previous administrative officer Gabi Klode. He approached me and said, "You are going back with me. We need you. Don't worry. It's all set. The Deputy Commander left our base and a new one replaced him named Menashe Inbar." I responded, "What? Menashe Inbar was my unit's commander when I had basic training." Officer Gabi then instructed me to follow him. I said, "But I have to first talk with the office sergeant!" Gabi said, "That's not necessary. He already knows your situation and everything is set." I happily agreed to accompany Officer Gabi.

As we returned to the base in Baloza, I saw the new Deputy Commander Inbar passing by in his jeep. He pulled over near and said, "Welcome back, Michael. How do you feel?" I responded, "I feel okay, sir." He told me that I would be the staff sergeant in the office and he was happy that I returned. I received a lot of confidence from the Deputy Commander's warm welcome and I was happy to serve again with friends that I knew. And above all, I have become the deputy administrative officer. I fulfilled my duty in the best way possible.

Order and discipline returned to the office after the negligence that happened while I was away. Gabi Klode had difficulties managing the office, especially controlling the female workers. First Sergeant Feldman helped me to discipline anyone who caused trouble with his special punishments. For more difficult problems, we wrote a complaint together to the Deputy Commander who judged and punished the troublemakers.

One of the punishments was not allowing the soldiers to take leave for their weekend vacation. This was considered a severe punishment and everyone feared it. That is how we fulfilled our duty for three months in the secondary front line in Baloza, and for another three months in the center of Sinai in Birt Madeh.

Two years had already passed since I was recruited and one more year was left for me to serve. During this period of time, I didn't encounter any extraordinary incidents. The administrative officer Gabi left our base and a new officer named Moti replaced him. He was a lieutenant who was shy and introverted. This job was too much for him, but I provided a lot of assistance as his deputy. One weekend I would travel home for vacation, and he would go the next.

One weekend, all of the officers went home for vacation and it was just the staff sergeant and I who remained on the base and were in charge. I was sitting in the office before the eve of Shabbat waiting to go for the celebratory dinner. It was winter time and the nights in the desert were very cold. Suddenly, the door opened and the sergeant in charge of communication burst in. He was very pale and afraid while he told me that the main communication equipment was on fire. I immediately left the office and saw that his small building was enveloped in flames that reached the sky. Some other soldiers joined to see this terrible sight and none of us knew what to do. I then said, "Let's run to the firefighting station and get sandbags and water buckets from there!" All of them accepted my suggestion but when we returned it did not help at all since it was impossible to get close to the building because of the flames. In the end, the whole building burned to the ground and only a pile of ashes remained.

On Sunday, a military investigative committee appeared to check the cause of this blaze. Their conclusion was that the communications sergeant had an electric heater and left it on when he exited the building. My date of discharge was getting closer and we again moved to the secondary front line in Baloza. Almost every weekend, I left the base on the flight that went to Tel Aviv from Sinai. On most of these flights, First Sergeant Feldman was on board and we became close friends. He was a very tough guy who was highly feared by all soldiers. It was unbelievable that we became friends. I knew this happened because of the flights we shared together.

The last night before my discharge in Baloza, I threw a discharge and separation party for my comrades. I personally invited the base commanders, particularly the deputy battalion commander, Menashe Inbar with whom I served during many periods of my military service in different places. He was admired as a professional commander by all of his soldiers and it was a great honor for me that he came to my leave party. He said to me as he made a toast, "I am sure you will remember us during your civilian life. I wish you a pleasant discharge and success in all of your future endeavors." Even though the party lasted well into the night, the deputy commander was in a good mood while he disregarded the difficult tasks that awaited him the next day.

The soldiers talked about the different adventures we experienced during our three years of service and particularly made fun of First Sergeant Feldman. At the time, I was sure that we would miss him a lot because of all the fun we had with him. When I write about him now, I see him standing in front of my eyes as a tough, strict, and sharp person who fulfilled his duty in accordance to the demands of the Israeli army. We former soldiers have to thank him for this, and have to thank all of our commanders in general who performed their duties in the best way, and taught us to be courageous and real men.

Now that it has been many years since I was discharged and I have a broad perspective of life, I see all the behavior of the commanders in a very positive manner. In my view, every Israeli who did not serve in the army has missed out for the rest of his life. I served, explored and wandered during my three years of service the whole Sinai Peninsula, and met different types of people. We together experienced many exciting moments as well as frightening ones. We literally ate the dust of the desert.

I would say that during my military service, I became friends with all the soldiers I served with. To this day, I cannot forget even one of them, and I am curious to know what fate befell each of my friends. After all of our discharges, we were absorbed back into civilian life.

My friends and I on draft day, February 5th, 1970.

Basic training, May 1970 in the Sinai desert.

Me, patrolling my bunker opposite Ismaeliya. Egyptian forces used to constantly bombard our bunker-camp with artillery shells. Just before the cease-fire I was wounded when an artillery shell exploded near me, injuring my lower back. Even with the injury, I remained on duty guarding my post for the next few days until my commander ordered me to visit the hospital. I returned to my base after 3 months.

During the cease fire of the 1970 conflict. My friends and I are enjoying some down time in the Suez Canal.

CHAPTER 23

It was a Monday in February 1973; a foggy morning which had no hint of glorious combat nor the tranquility of ceasefire. It was a tired morning almost like the desert itself. There was nothing special that would have been appropriate for my military discharge. That day had arrived, a day that was difficult to forget. I took a last glance at the canal and its surroundings. I was wondering if I would ever return. Secretary Rogers' peace plan among Israel, Egypt, and Jordan was underway and there would be peace if Israel would retreat from most of the conquered lands. If this plan would be fully implemented, the armored brigade whom I belonged to would soon leave the canal line; like I did that day. This brigade was well trained and served consistently in the Sinai Desert since the Six Day War despite many casualties.

On the morning of February fifth, I left the Hatzav military camp with my duffel bag containing my olive uniform that served me during my entire military career. I went to the airport in Baloza for my flight to Tel Aviv. Upon boarding the plane, I thought to myself that maybe this flight would be the last time that I would get to see the sands and mountains of Sinai. I then thought that maybe the army would call me back as a reservist to serve there. This would not have been likely in the near future.

While the plane flew over the sands of the Sinai Desert, I was overcome with the good feeling that I was going to be a civilian again. After the plane landed in Tel Aviv, I went to the recruiting station to return my equipment and uniform and receive an official discharge certificate. I was very happy to meet many of my friends who I was recruited with three years ago there. Each one of us served in a different place. At last, we were together again and had the opportunity to share our stories.

As a recently discharged soldier, the department for veterans in the Ministry of Defense immediately found work for me in the aviation industry. I worked in an electro-chemical unit, which was a very interesting experience. My relations with my coworkers were positive because they respected me as a fresh veteran and the youngest among them. I lived near the plant and did not need to use transportation.

Since I had returned to civilian life, the Egyptians had threatened from time to time to start another big war with Israel. The Israeli people as well as its government did not care about these threats. The tremendous victory of the Six Day War and the sudden death of Egyptian president Natzer inspired a feeling that it would take a long time until the Arabs would be ready for war again.

In June 1973, four months after my discharge from mandatory service, I received a recruitment summons. It briefly stated that my first service as a reservist would be in July for two weeks for the purpose of establishing a new battalion. As I arrived in the military training camp, in Southern Israel at the Negev Desert, I met the administrative officer of the battalion, Lieutenant Menachem. After an interview in which I said that I have over a year-and-a-half's worth of experience as a sergeant in a military administrative office, he decided to appoint me as his aide. We immediately started to absorb new recruits. I knew very few of them from my previous mandatory service.

Within two weeks we succeeded to establish a battalion composed of four platoons: tanks, armored vehicles, jeeps, and infantry under the command Lieutenant Colonel Bentzi Carmel. This battalion was attached to the military division of General Ariel Sharon as a reconnaissance battalion. There was absolute tranquility in the land of Israel and no one expected a war in the near future. However, the Israeli leadership knew well that the Arab forces, particularly those of Egypt and Syria, were preparing themselves for war.

From 1971 until May 1973, the Egyptian army called back many of their reservists three times and Israel responded by calling back their own reservists. Consequently, nothing happened and Israel came to the conclusion that Egypt was trying to engage in a war of nerves. For Israel, this recruitment caused tremendous economical damages since all of these reservists had to leave their places of work to return to the army. When Egypt called back its reservists for a fourth time during the Fall of 1973, Israeli intelligence thought it was just another ploy and recommended to the high military command not to follow Egypt's example. This reminded me of the story of the boy who cried wolf. When the threat was real, no one would have taken it seriously.

At the beginning of October 1973, some of those who were observing the Suez Canal reported to their commanders that they had noticed Egyptian troops planting pegs in the water. They added that in their assessment, the pegs were meant to have a bridge built over them. Since this assessment came from those of low rank, it was not given appropriate consideration.

At my place of work in the chemical department of the aviation industry, the manager gathered together all of the employees on the fifth of October to wish us an easy fast on *Yom Kippur*. The evening of that day, was the start of *Yom Kippur* fast. We were dismissed from work in the early afternoon in order to prepare ourselves for the upcoming holiday. This day is considered the holiest in the Jewish religion and it lasts for twenty-four straight hours from evening until the evening of the next day. The state of Israel is basically a secular country, but *Yom Kippur* is considered the only holiday when all offices and official institutes as well as private businesses close in accordance to state law.

During these early afternoon hours, all roads are jammed with traffic by commuters and drivers who are making their way home. The day after, the roads are void of vehicles. Only physicians and military personnel on duty dare to violate the sanctity of this day and travel by vehicle. On this day, even the media is paralyzed; there is no radio, TV broadcasts, or newspaper publishing. Besides fasting, the people spend most of the day praying in the synagogues to ask forgiveness for the sins they committed the whole previous year. It is interesting to see that even among secular Jews, many observe this holiday very strictly. The rest usually do not leave their homes like they use to do on other holidays when they would spend their time having picnics.

This silence and tranquility during this extraordinary holiday inspires a unique atmosphere of holiness over the whole land of Israel on *Yom Kippur*. On the evening of that *Yom Kippur*, half an hour before the start of prayers, the central synagogue of Yahud where I lived was totally crowded with men, women, and children. Many from the crowd were standing outside the synagogue just to meet and talk with acquaintances. I also stood outside wearing a Tallit like the other men and talked with my friends. This was a smooth conversation in which we did not discuss the current tense military situation but only talked about the fast and common friends that we knew. The main street of the town was quiet and empty of traffic. Only some late-comers were seen there rushing towards the synagogue.

For the young men and women, this was a unique occasion to meet and talk. Someone asked me, "Where is your brother Ishai? Is he stuck in his military base?" I answered "Yes, he is trying the best to earn a vacation for the upcoming *Sukkos* holiday." We then saw David, one of our friends approach us along with his girlfriend. "Hi guys. What is going on?" he asked. "Did we come here to pray or meet? I did not want to come but my girlfriend dragged me with her." Then David began to boast about his successes as a soccer goalie. "Do you know what accolade I received? I've been the best soccer player on the field. As of now, no one has ever been able to score a goal on me. Since I joined the team this season, we have had only victories and very soon we will compete for the national cup against Israel's number one soccer team, Macabi Tel Aviv." He was not so far from reality as we could read in the newspapers about his successful performances.

When he finished talking about soccer, he asked, "Do you plan to be in the synagogue all day tomorrow? Maybe we can go to someone's house to have a chat." Then someone answered, "Tomorrow we will sleep as late in the morning as possible, and then go to the synagogue for a few hours; in the afternoon we will indeed meet up at one of our homes." David responded, "It sounds good because this way we won't suffer too much from fasting." Then another friend named Nissim arrived and someone from the group said to him, "Oh, I am happy you are here. Do us a favor and open your coffee house tomorrow specially for us. We will come and play chess and dominoes there to pass the time." Nissim responded, "No way. Tomorrow is the holiest day for the Jewish people and we will come to pray at the synagogue so we can have a successful year."

He wished us luck, left our group, and entered the synagogue. David then said, "Do you guys know that Nissim has a heart of gold? He is like our father. He takes care of all of us like we were his sons."

Suddenly, we heard the servant of the synagogue call out, "Everyone, be quiet and listen to *Kol Nidreh.*" The door and windows of the synagogue were wide open so even the people standing outside could participate in the ceremony. There were no vacant seats in the synagogue, but my father took care of me and had reserved a seat next to him. I immediately entered and sat down. The cantor then started *Kol Nidreh* and after a prayer of over two hours, the congregation left the synagogue.

Afterwards, all of my friends gathered together to wander the streets of Yahud. Many other couples, friends and families were doing the same. The main street was glittered with the lights from the stores and people were wandering dressed in fancy clothing for the holiday. No vehicle was seen or heard in the streets. One by one, our group separated as we wished each other good night and to meet again in the synagogue the next day.

At that time, I was seeing my first girlfriend who also attended the prayer service. After I separated from my friends, I accompanied her to a private home that was close to mine where she had a babysitting job. Late at night, I left that home and I agreed with my girlfriend to meet again tomorrow evening at the end of the fast to tour the city on my scooter.

As I left the building, I suddenly saw a military truck that had just parked and people from the neighborhood were surrounding it. I was wondering what it was doing on our street on the eve of *Yom Kippur*. I approached the vehicle and the driver got out with a recruitment order in his hand. He asked us, "Could you help me find a soldier who lives in this neighborhood?" Someone from the crowd gave him directions to the soldier's house and he left. At that time, an argument started among the people surrounding the truck. "What is going on with this army if they would summon a reservist on the eve of *Yom Kippur*? Could they not wait until tomorrow evening?" Another person said, "People, isn't this an army? They may need him urgently." At that time, nobody there could imagine that near the Israeli borders, there was unusual enemy activity. A few minutes later, the gathering near the truck scattered and I went home to sleep.

The next day was *Yom Kippur* and I woke up early in the morning. I dressed myself and left home to the synagogue which was a very short distance away. It was a pleasant and quiet morning. People were seen in the streets rushing to the synagogues. In the distance, I could hear the voices of the cantors who were asking forgiveness for all sins committed during the previous year. When I arrived at the synagogue of my congregation, it was already full. Those who had not reserved a seat in advance had to stand in the hallways or in the outdoor yard to attend the service. I sat in my reserved seat next to my father.

Upon my arrival to the synagogue, rumors were spread among the congregants that a calling back of reservists was taking place and the Israeli army was in a state of alert. A few military vehicles were seen going quickly down the street but no one in the synagogue knew exactly what was going on. From time to time, Dad said to me that maybe this was a recruitment exercise. I did not accept this opinion, since such an exercise would not have taken place on the holiest day of the year unless there was a serious emergency.

Suddenly, an unfamiliar man entered the synagogue and approached the cantor. After a minute of quiet talk, the cantor took off his *Tallit* and quickly left the synagogue. At that minute, I knew that Dad's recruitment exercise theory was wrong and that war had erupted. Dad still insisted that he was right.

I could not sit calmly and I left the synagogue to its yard. I wanted to hear from the people there what was going on since it was forbidden to freely talk inside the synagogue. All the information that I received from them were just assessments. I then decided to go home. On my way, I met the father of my friend Leon. He told me that a few minutes ago, Leon was called back from reserve. Afterwards, I met my friend David Cohen, who served with me in the same armored battalion. He was standing in front of his home and he said to me, "Be ready to be called back since a recruitment exercise is in progress." I instantly responded that I did not think so, rather we are really in a stage of war and that I was almost certain of this.

I continued to march towards my home where many children were playing out front. My little brother Avi saw me and ran towards me saying, "There is a military alert and many reservists from the neighborhood have been called back to the army." The time was a little before two o' clock in the afternoon, so I said to Avi, "Let's go home and rest." I had hardly gotten into bed when my brother rushed in and said, "A person with a recruitment summons is looking for you, out in his car." While I was quickly getting dressed in my synagogue clothes, the man arrived at my house. He told me to join him in his car so I can be taken to an assembly point for soldiers. I remember myself asking him, "Is something serious going on? Do I have to take with me any special luggage?" He answered, "No. It is just a regular exercise and you will be back in the evening."

At that moment, a loud siren was heard at exactly two o' clock. Mom didn't waste time and immediately prepared a few sandwiches for my journey. The recruiter rushed me to leave with him since it was late and he had to pick up the battalion commander who lived in Savion.

As I entered his car, I asked him to turn on the radio on the Israeli military frequency. I was nervous and under stress. Nevertheless, the beautiful song of the famous Israel singer Chava Alberstein, *Eretz'Ahovati* calmed me down. Suddenly, I heard the radio commentator say, "Around two o' clock, the Egyptian and Syrian forces simultaneously opened fire all over the borders." The truth is that I was not surprised by this announcement but felt that I, as well as all the other young soldiers, was obligated to reach the frontlines as fast as possible. Little by little, the roads became crowded with vehicles which were summoned by the army to collect reservists from their homes. Israeli military airplanes were frequently passing over our heads. This sight only heightened our stress.

On my way to the assembly point, I saw recruiters entering synagogues and extracting reservists. Through the windows of houses, I could see wives and mothers waving goodbye to their loved ones who left their homes dressed in uniform. The radio repeated the same news over and over. After an hour's drive, we reached the assembly point. Dozens of buses were waiting there to pick up the crowds of reservists. I, dressed in my civilian clothes, felt uncomfortable and wanted to go back home and get my uniform and equipment. Seeing these buses with their engines running, I decided to travel to my military base in the Israeli Negev desert and ask for new equipment and a uniform. Two hours later, we reached the camp. The administrative officer of the battalion, Lieutenant Menachem, was expecting my arrival and immediately instructed me to start registering the returning reservists. I asked him if the officer in charge succeeded to deliver recruitment summonses to all the battalion troops and he responded positively.

I said, "If this is so, we don't need to waste time on administrative matters since every soldier knows which unit he belongs to as the battalion has recently been established." Menachem did not agree with my view. Another soldier and I started to register all the reservists that arrived until midnight. We soon received reports from the bunkers at the Suez Canal of terrible battles between the Israeli and Egyptian forces. Mass bombing from the air and from artillery was used by both sides. The worst news of all was that Egyptian forces had started to cross the canal.

Israeli intelligence reported that five infantry divisions, three vehicular divisions and two armored divisions of the Egyptian army were deployed close to the Suez Canal. Each division that intended to cross the canal was reinforced by an armored brigade. Simultaneously, two thousand artillery mortars opened fire along the borderline and turned the whole eastern bank of the canal into a hell on earth. At the "official" start of Egyptian aggression they sent 240 airplanes to cross the canal. At 2:15 p.m., as they ceased their bombardment operation, the first wave of 8,000 Egyptian infantry troops started to cross the canal. Most of these invaders preferred to flank the Israeli bunkers and advance to the east. They were ordered to create an Egyptian enclave two to three kilometers away from the canal. Their mission was accomplished by sundown.

The flow of reservists to our base seemed endless. Their registration process prevented them from being quickly equipped and moved to the frontline. Surprisingly, General Ariel Sharon accompanied by the battalion commander, Bentzi Carmel, appeared at the base at midnight of *Yom Kippur*. The general noticed the hindrances that resulted from the registration process. He immediately gave an order to cease all administrative processes, get the tanks ready, and move to the front. His division was deployed to the central zone of the Suez Canal line.

Upon receipt of the general's order, soldiers boarded tanks, armored vehicles, and jeeps and waited for the order to advance to the front. The others rushed to the armory to acquire submachine guns. I also asked for a military uniform and a helmet. All of this was an unbelievable scene. Reservists who until this day were complacent civilians at their homes with their families or attending *Yom Kippur* services in the synagogues, had found themselves within a few hours in a chaotic situation.

By four a.m. I left the base with the last of the soldiers. This was a long convoy of combat forces. The administrative officer Menachem and I drove along in a civilian van that was overloaded with MRE's, maps, and other supplies. We moved through Be'er

Sheva and the capital city of the Negev towards Sinai.

The main street of the city was crowded with women, children, and the elderly. They gave us free newspapers, coffee, and food. It was a spectacular view to see Jewish women of Middle-Eastern origin raising their arms to the sky and blessing the fighters upon their departure to the battlefield.

CHAPTER 24

In July of 1973, General Shmuel Gonen replaced General Sharon as the chief commander of the Southern zone. This zone included the southern part of Israel from the city of Be'er Sheba via the Negev desert to the port of Eilat at the Red Sea. It also included the whole Sinai Peninsula, which is a bridge between Asia and Africa. General Gonen, known by his nickname Gorodish, was a very tough military man who was feared by his own soldiers. The general who was in charge of protecting the Suez Canal line was Abraham Mendler, nicknamed Albert. In contrast to Gorodish who was his commander, Albert was one of the friendliest commanders in the Israeli army. The forces that were under his command at the outbreak of the war included 280 tanks in three brigades and one infantry brigade. He was scheduled to complete his duty on Sunday the seventh of October and be replaced by General Kalman Magen.

The whole week before that day, Albert would visit different units of his forces to bid them farewell. On Saturday the sixth of October, which was *Yom Kippur*, in the early afternoon, it was clear to Albert that the Egyptians had initiated a total war against Israel. Around four o' clock, the Egyptian forces crossed the entire length of the canal in a massive amphibious assault. The Israeli defense line at the canal consisted of the maozim bunkers that had a distance of ten kilometers between them. In these ten kilometer gaps, the Egyptians constructed temporary bridges in which their forces crossed the canal. They advanced quickly into the Sinai Peninsula without coming into contact with the bunkers. The result of this strategic move was that all the troops in the maozim were surrounded on all sides by the Egyptians.

The armored brigade of Colonel Amnon was stationed in Sinai a few kilometers away from the canal. This brigade was well-trained and had served persistently in Sinai since the Six Day War. Colonel Amnon was a tall man with a wide mustache and glasses. He was the commander of this brigade for a year so far. I had been discharged from one of the battalions that belonged to the brigade of Colonel Amnon just a few months prior. I knew most of the commanders and troops of this battalion well. When the Egyptians started the massive bombardment, Amnon's troops boarded their tanks, moved towards the enemy, and returned fire. The losses and casualties of this brigade were tremendous during the first two days of war.

On Sunday evening, the second day of war, General Sharon had arrived on the scene. He immediately gave an order to evacuate the troops from the maozim. The soldiers there who had heard his voice and the order had their confidence renewed. In the communication system, I could hear them saying, "We are now sure that everything will be alright. We were afraid and concerned but now that we have word of Sharon's arrival, we know that other forces will come to our rescue." This important order was consistent with General Sharon's military philosophy to never abandon his troops regardless of their condition on the battlefield.

That evening, our supply unit reached Camp Tasa and settled there. The combat forces of our battalion were many kilometers ahead of us and we did not know if they were in combat or not. The next day, our unit commander, Captain Shavit, met General Sharon who was accompanied by the Chief Chaplain of the army, Rabbi Goren. They drove around the zone in a jeep to assess the situation. General Sharon told Captain Shavit that our battalion had engaged combat and they urgently had to be supplied. We immediately organized supplies of food, water, and gasoline. We then loaded them onto trucks and sent them over to our fighting forces.

The supply unit stayed there and felt relieved. We even mistakenly assessed that this war was a repeat of the Six Day War. And as two days of combat had elapsed, we believed that the war would soon be over. We did not imagine that this time it was a totally different war.

The first detailed reports came from the front line and they were very discouraging. Bitter fights between large forces took place and the casualties on both sides were tremendous. The commander of our battalion, Lieutenant Colonel Bentzi Carmel, got killed in action and many of our tanks were destroyed and our battalion retreated back. This report was the worst for my friends and I. I personally could not believe that Lieutenant Colonel Bentzi, who was a cool commander with a lot of battle experience, got killed in such an early stage of the war.

In the Six Day War, he was one of the most successful commanders and was even seriously wounded at that conflict. His death was a shock to the whole battalion. He was actually the first soldier of General Sharon's division who was killed in action. The general gave the order to retreat in order to reorganize the battalion under a new commander. He appointed Major Yoav Brom to replace Carmel. Major Brom was a member of a *kibbutz* and a young, compact man.

The Tasa camp soon turned to be chaotic quite unlike the quiet days when I served there during my mandatory service. The tank hangar of the camp was getting filled with damaged tanks and other vehicles that needed to be fixed. The huge bunker that had just recently finished construction was turned into a field hospital. Many severely wounded soldiers who were brought from the canal line laid on stretchers and cried in pain. Only one physician was able to take care of all of them and his name was Doctor Adler. At the entrance of the bunker, I saw piles of shoes and bloodied socks. I wondered if they belonged to dead soldiers.

Our armored brigade, who retreated after the death of its commander, stayed a few kilometers south of Camp Tasa. There, they waited a few hours for the order to make a counter-attack. When the order came and the convoy started to move towards the canal, we suddenly heard a loud alarm. It was accompanied with an announcement over an address system, "Be prepared for an air-borne assault." From that time on, that announcement became routine.

Many troops gathered near the bunker and someone shouted, "Here are the enemy planes!" This was the first attack by Iraqi combat planes that fought in cooperation with the Egyptian forces.

They did not strike Camp Tasa itself but their actual target was the armored convoy. Anti-aircraft artillery shot down one of these planes. Camp Tasa was turned into a gathering point for soldiers. I met many of my old friends from both my mandatory service and civilian life there. Some of them stayed there with no duty or purpose. One of them was Eli Cohen from my neighborhood in Yahud. He was just sitting on a truck and reading a newspaper. I greeted him and asked how he was doing and what his plans were. He did not know what to answer me. He said impatiently, "We were brought here as truck drivers and for two days, we were assigned no duties. We were bored and from time to time, we ran to the trenches whenever the enemy planes flew by for another assault. When will all of this end and when will I be assigned to do something?"

Another friend from Yahud, Moshe Vajeema, was not complaining since he had a duty from the start of the war. He was assigned to evacuate the wounded and transport them over to the bunker which was turned into a makeshift hospital. Another friend, named Rappaport, was a supply truck driver and whenever he saw me he asked for news from the frontline and was eager to know when the Israeli army will cross the canal to the Egyptian side and continue the fight there. More old friends of mine, Sasson and Nachesi, with whom I served during my mandatory service and were very close with each other, were on duty together. They took care in advance to serve together as reservists since they had already been close friends as civilians.

My friends were very brave soldiers and were eager to get a combat assignment. We were always on alert for Egyptian commando raids. My friends and I were assigned to guard the camp in turns. The eighth of October, the third day of war, was very critical. While the division of General Sharon fought in the central zone of the canal line, another division led by General Bren, fought in the northern zone. In the early morning, General Bren received the final approval to start a counter attack with his three brigades from north to south. The purpose of this attack was to destroy the Egyptian forces that invaded the eastern bank of the canal, thus enabling the Israeli forces to cross the canal westward.

This counter-attack failed and General Bren's division suffered many casualties and even lost important positions. On that day, our division was stationed in the central zone and wasted time. General Sharon became impatient and did not want to wait until the other division would succeed in their mission to destroy the Egyptian forces north of us. He decided to launch an attack on the secondary front line which had already fallen to the Egyptians. He carried out this assault with two brigades and with no approval by the high commander of the Southern zone, Major General Gorodish, who was above him in the chain of command. From that day on, generals Gorodish and Sharon, had a bitter relationship.

The Israeli government understood that such a relationship between high commanders was a liability to the management of the war. They appointed Lieutenant General Reserve Bar Lev as the personal representative of the chief of staff for the Southern zone command. In actuality, this move was a ploy to remove Gorodish from his position. At that time, Bar Lev had a civilian position as the minister of trade. He was summoned to active duty on Wednesday October tenth.

The Egyptian fighter planes bombarded the Israeli forces, especially at daybreak. The Israeli Air Force did not intervene at all since all of their planes were busy with managing the Northern front against Syria. This was a strategic decision by the Israeli high command since it was critical to push back the Syrian Air Force from the Israeli towns in the Galil. When they finished this mission, they moved to the Egyptian front.

On the early morning of October ninth, Camp Tasa and its surrounding artillery units were assaulted by Egyptian fighter planes. Most of our soldiers were still sleeping either in their vehicles or on the sand. Lieutenant Menachem and I slept in the civilian van. Menachem was a lawyer in his civilian life as well as devoutly religious. His primary weapon of choice was his *Tallit* and *Tfillin*. Every early morning, he wore them during his morning prayer and he passed them over to me when he finished. On that morning, he had barely opened his eyes from sleep when he was thrown out of the van by the shockwave of the assault. He shouted orders for us to spread out to minimize casualties. There was a standing order that immediately at daybreak, all of the soldiers had to scatter; but unfortunately, this particular morning the rule was not complied with. The consequences were terrible and many soldiers were killed or wounded.

Before sundown, Menachem and I decided to travel to Refidim to find entertainers who were willing to go back with us to the camp to recover our men's morale. When we returned with them, the battalion commander lost his temper and shouted, "Who asked you to go to Refidim? We are expecting an Egyptian commando strike and all of you should scatter immediately and find shelter. The entertainers, who brought their guitars with them, took cover in the tanks while Menachem found shelter in one of the armored vehicles and I did the same in another.

The operator of this vehicle was an old friend of mine from my mandatory service, Lieutenant Tenenbaum. He immediately assigned me a position to be his shooter in case we encountered the enemy. The night passed with no incident. Lieutenant Tenenbaum and I spent this night by exchanging stories of our experiences during mandatory service. Towards daybreak, the battalion commander announced over the address system to prepare ourselves to move out. He also ordered Menachem to arrange the safe return of the civilian entertainers to Refidim. Menachem was very disappointed that the entertainers didn't have a chance to perform.

Plenty of soldiers were waiting in the camp and its surroundings and had no idea what was to come. There was no clear plan of action. They did not even know what to do with the vehicles which were damaged. We listened to the radio announcements of the military spokesman. He was the only source of information for us to have a rough idea of the events in the battlefield. We still hoped that like in the war of 1967, we would soon hear from our spokesman good news of victories and miracles. This hope was not realized.

That day, Menachem and I took the initiative to shower for the first time since *Yom Kippur*. On the fourth and fifth day of the war, rumors spread among the soldiers that the military high command was preparing plans to cross the canal for an incursion into the Egyptian mainland. On one occasion, I heard directly from the battalion commander, Major Yoram Brom, that he was actually waiting for an upcoming order to cross the canal. I then saw the renowned television journalist for military affairs, Ron Ben Ishai accompanied by a cameraman. He was reporting for the Israeli T.V. the events of the battlefield.

On Saturday, October thirteenth, the Israeli military was dealt a severe blow when the distinguished commander, General Albert, was killed in action. Albert was standing in his armored vehicle, observing enemy movements through a pair of binoculars when a missile struck his vehicle and killed him instantly. On that day, ground-to-ground missiles flew over Tasa and ballistics of all kinds were being shot from all directions. Some friends of mine succumbed to panic and were desperately wondering how this would all end.

That Shabbat, Menachem organized a mass outdoor prayer and many attended this service. Both religious and non-religious soldiers joined in this prayer to beg G-d to relieve them of despair and grant victory.

On the early morning of Sunday, October fourteenth, Egyptian tanks launched a massive assault on the Israeli forces. Commentators described this event as the largest tank battle in history. About 2000 tanks from both sides were involved in this epic battle. The Israeli army was expecting to win this decisive battle, fully aware that only after achieving victory could they attempt an incursion across the Suez Canal.

Captain Rafi Bar-Lev, cousin of Lieutenant General Chaim Bar-Lev, was abroad when war erupted. He immediately returned to Israel and joined the battlefield. He was the commander of a tank unit in our battalion. On that day, he and his soldiers succeeded to destroy the enemy tanks one after the other. At the end of the battle, Captain Rafi was very satisfied with the results and was heard in the communications system saying, "If we cannot celebrate our *Sukkos* holiday now, we at least had a nice *Lag Ba'Omer*." *Lag Ba'Omer* is a Jewish festivity involving bonfires.

This crucial battle ended with a decisive victory for the Israeli forces. Hundreds of Egyptian tanks were destroyed or abandoned and our morale skyrocketed. The Egyptian's ignominious defeat was a wonderful prelude for the upcoming Israeli incursion across the canal. Our division under the leadership of General Sharon received the order to initiate and lead the crossing.

On Monday, October fifteenth, the Israeli Air Force launched a massive bombardment on the eastern bank of the Suez Canal, at the point where the incursion would take place. The purpose of this air strike was to destroy the Egyptian's anti-aircraft artillery at that location. When this mission was accomplished, the crossing of the canal began. A brigade of paratroopers from General Sharon's division used inflatable rubber boats to cross the canal. This brigade was accompanied by a unit of ten tanks carried on rafts. These soldiers had to carry out the task of securing the western side of the temporary bridge that the civil engineer unit hastily built. The path, which was composed of several parallel routes, between Tasa and this bridge was jammed with all types of military vehicles: tanks, armored vehicles, jeeps, and supply trucks. They all made their way south en route to the bridge.

I had the feeling like the whole country was going south. On the way, I met plenty of soldiers who were long-time friends. They were filled with confidence while riding in their military vehicles. They swiftly passed as they shouted, "See you in Egypt!" The loud roar of the Israeli airplanes could be heard above us throughout the day as they flew to the west side of the canal to drop their bombs and return to be re-armed.

Military journalists, television staff, and cameramen were also seen on this path traveling with the fighting forces. They filmed and described the advancement towards the crossing. I said to Menachem,

"If the Egyptians knew what was coming they would be miserable now." Menachem instantly responded, "Why miserable? They don't even deserve to be alive for what they have done to us." Menachem was reciting phrases from the book of Psalms.

Each soldier received a copy of the book of Psalms from the chief military chaplain, Rabbi Goren before heading out. From time to time, Menachem would say to me, "Turn to a page and read it loudly." At that time, we had received an order from Captain Shavit who was the commander of our convoy to pull over to the side of the road and enable the combat fighters to move ahead. Menachem and I fulfilled this order and observed those who passed us as well as the planes that continued to fly overhead. As we continued to wait on the side of the road, two combat soldiers approached us and I immediately recognized them as Moshe and Avi, two friends of mine from Jerusalem. They were traveling with our convoy since they had lost their own unit during the first days of fighting when their armored vehicle was put out of commission. Now they were begging Menachem to assign them to one of our battalion infantry units.

Menachem promised them that if there were any vacancies, they would fill them. Later on that day, they boarded on an armored vehicle that had space for them in accordance to Menachem's referral.

The advancement to the canal was interrupted by frequent ambushes by pockets of Egyptian forces on the eastern side that attacked on all routes to the canal. The Egyptians stubbornly fought and opened heavy fire from all directions. Our brigade led by Colonel Amnon lost many good men, and my battalion commander Major Yoav Brom. Clearing the routes from the Egyptian ambushes took a long time, as did taking care of the wounded. Chaos had erupted on all routes. At one stage, it took our convoy two hours just to advance seven kilometers. Some of our ammunition in the supply trucks exploded due to the attacks causing many casualties and many armored vehicles were taken out of commission.

On this terrible journey, our brigade had been severely stricken and lost more than half of its original personnel. My own battalion had remained without a father figure since Major Brom had been killed. At some point, Captain Shavit, the commander of our unit, gave us an order to stop and go the opposite direction. We were shocked to hear this order but had no choice but to comply. I said to Menachem, "What is going on now?" He answered that he had no idea. "I think that our commander Shavit has no confidence. All of the combat forces have moved forward, and us at the end of the convoy, he ordered to go back. I cannot understand this bizarre order to miss the historical cross of the canal."

Menachem then said that an order is an order and that he will enquire the rationale for it when we are back in Tasa. It was before midnight when we reached Tasa. Everything there was quiet and the roads were vacant after all those traffic jams a short while ago. No noises of moving tanks, airplanes, or artillery were heard there. As all of the war machines continued to the south to carry out the courageous mission. Everyone who had returned waited to receive a report from Shavit for the reason of this return.

We were very frustrated and asked each other why we had to return. Some of my friends turned on the radio to listen to the news of the canal crossing. Suddenly, the news was interrupted with a song and a blessing to a soldier who had just became a father. One of my friends jumped with joy yelling, "That's me. I am a father. I have a son now." The rest of us wished him *Mazal Tov* and we returned to contemplate our problem.

Menachem went to talk with Captain Shavit and when he returned, the problem had been solved. Shavit explained to Menachem that our division no longer existed. The commander of this battalion, Major Yoav Brom, was killed. He was our second battalion commander to be killed in action. Many of our commanders and troops were wounded as well. Menachem continued enthusiastically with his explanation. He then said, "We are going to collect all of the remaining troops of our battalion who had been scattered to other units, to make a list of our dead and of our casualties."

On Tuesday, the sixteenth of October, Menachem and I went to visit the field hospital in Tasa. There we received information about casualties from our unit who were referred to different hospitals in

Israel. With this information, we were able to make a final list of our casualties and the troops who were still with us. The remaining troops of our battalion who had made a U-turn with Captain Shavit moved on to Refidim and waited there for new orders. They wandered there with no purpose while the other remaining troops of this brigade joined various units of other battalions.

Menachem and I continued to Refidim to visit our troops. We found them listening to a direct broadcast from the radio of a government session. Prime Minister Golda Meir was the first to make an address. Afterwards was a speech from Minister Menachem Begin who had just joined the government due to the emergency situation. Until then, he was the leader of the opposition in the Israeli parliament and by joining the government, he could attend its sessions despite him being a minister without a portfolio.

We listened carefully to his enthusiastic speech and many of us were moved to tears. For the first time, he mentioned the activities of the Israeli forces on the western side of the canal. He revealed that the Israeli artillery, along with the air force, had thoroughly bombarded the entry point for the troops on the opposite side of the canal in order to prepare a safe corridor for the Israeli army.

From that day on, Menachem and I traveled between Tasa and Refidim. Occasionally, we went to the frontline to receive reports of the activity of our brigade of troops on the other side of the canal. Once we reached the foot of the canal bridge in an armored vehicle, and a terrible Egyptian air strike occurred. I suddenly heard someone screaming, "Oh, my foot! Oh, my hand!" I got out of the vehicle and ran towards this soldier. I could not reach him since at that moment four Egyptian planes flew overhead and dropped napalm on us. I immediately lied on the ground and saw the whole surrounding area burning. Vehicles exploded and many soldiers were wounded or dying all around me. I comforted the wounded and encouraged them to not sink into despair until the paramedics arrived.

Even today, I do not understand how I survived that hell. It was a true miracle. Anyone who survived should have thanked G-d for being able to get out alive.

An interesting end to this terrible event was that two of our own planes appeared and within a few seconds. They destroyed three of the Egyptian planes while the last one escaped. This was a splendid sight for all the soldiers on the ground. A funny thing then happened. An Egyptian chopper appeared at that time in an attempt to make an incursion into the east of the canal. Almost immediately, ballistics of all kind were fired directly at this chopper and the pilot did not have as much luck as the one in the escaped plane.

One of the significant difficulties our forces faced was the evacuation of casualties from the western side of the canal. It was impossible to use the narrow bridge to transfer them to the eastern side. Therefore, the casualties were put on inflatable boats and were floated across the canal to receive immediate medical treatment on the eastern bank.

The division of General Sharon was advancing deep into the mainland of Egypt accompanied by heavy battles. Another division led by General Bren followed the division of Sharon by crossing the bridge and entering into the western side of the canal. Now the initiative was on our side. It was obvious that the Israeli forces could advance into Cairo, the capital of Egypt. Only outside intervention of the two superpowers, the Soviet Union and the U.S.A. could halt this advancement.

During these days, the most famous personality in our region was U.S. Secretary of State Henry Kissinger. He shuttled between Moscow and Tel Aviv for mediation. I remember a funny story from those days. I met a soldier who was accidentally wounded while repairing a tank in Tasa. I took him to the division's physician in the field hospital in Refidim. After waiting a few minutes in front of the physician's tent, I went inside with the wounded and explained to the physician what happened. "This guy, who will not reveal his name, was not wounded on the field but must be treated urgently. He claims to suffer from terrible pains in his left arm and it seems that a bone might be broken." After a short exam, the doctor decided to refer him to a specialist.

When he started to write a referral letter he insisted the wounded for his name. The soldier identified himself as Kissinger. The doctor said, "I do not have time to joke. Please give me your name." The soldier repeated the name Kissinger and added, "How come you don't believe me? You think there is only one Kissinger in the world? I am a Kissinger too." The doctor decided to refer him to a psychiatrist. The soldier started to yell and scream, "It is really my name and you will write it down!" When I returned to the car by myself, I told Menachem the funny event that I had just experienced. Menachem burst out laughing and said, "Seriously speaking, even I would have sent him to a psychiatrist."

Meanwhile, the whole state of Israel was talking about the more well-known Kissinger as if he was the messiah who would mediate a ceasefire agreement. I said, "Who needs a ceasefire now? The Israeli forces are advancing towards Cairo and within a day or two, they will defeat the Egyptian army decisively. Kissinger will prevent a great victory."

Menachem responded, "You don't understand at all how much a ceasefire agreement is needed. I think that Russia is going to get involved in this war right now to prevent Egypt's defeat. If this happens, this will be the war between *Gog* and *Magog*." I asked him, "What is that?" Menachem responded in amazement, "You didn't learn the Bible? It is written in the book of the Prophet Yecheskel, chapter 38, about the war which will take place in the Middle East between superpowers. All the evil people will die and the righteous will win and remain. In my view, all of what is going on now is similar to the description of the Prophet." So I asked him, "Then who is righteous and who is evil?" Menachem disregarded my question but many soldiers who had been listening were in fear of what the 'prophet' Menachem was saying.

Menachem reiterated this speech over and over until he got an encouraging feedback from the news broadcasted over the Israeli radio. Russia had declared a state of alert among its entire military force. The Russian president Kosygin warned that if Israel wouldn't stop its advance into the Egyptian mainland, he will deploy his forces to the area. Menachem boasted, "You see I am right! All of this is clearly written in the Bible… I did not make it up."

Another division of the Israeli army was led by General Kalman Magen, who had replaced the deceased General Albert. He joined the divisions of General Sharon and General Bren in the fighting in Egypt's mainland. All three of these divisions pounded the Egyptian army mercilessly and Russia started to send a squadron of cargo planes carrying all types of ammunition to the Middle East. When the U.S. president Richard Nixon received this news, he was outraged and he immediately ordered an alert for all American forces all over the world. American airborne units in several European countries got ready for the order to deploy. During those moments, it really seemed that the globe was on the verge of World War III. Of course, Menachem reiterated his 'prophecy' again and said, "Guys! This will be the real war!"

On October twenty-second, the Israeli radio broadcasted the news that the United States and the Soviet Union had offered together a ceasefire proposal to the two parties involved in the fights. It was planned to go into effect at seven p.m. that evening. I asked Menachem for his opinion about this news. "Do you think the ceasefire will go into effect? Would the Israeli government be ready to accept this proposal since the Israeli army was currently prevailing over the enemy?" Menachem answered that in his view, the Israeli soldiers were as exhausted as the Egyptians. "We lost a lot of soldiers and equipment."

It was already clear that the whole Egyptian army was on the verge of collapse. The two opposing parties were indeed exhausted but the Israeli forces had more breathing room and could surround the Egyptian forces. The divisions of Bren and Magen threatened to occupy the city of Suez, while Sharon's division threatened to occupy the city of Ismaeliya. General Sharon made maximum efforts to make last-minute achievements on the battlefield before the ceasefire agreement went into effect. Minutes before the scheduled time of the ceasefire, the Egyptians delivered a heavy artillery bombardment on the Israeli forces which lasted for about fifteen minutes. This was a massive and terrible attack which caused many deaths and casualties on the Israeli side.

From all around, I heard the moaning of the wounded crying for help. At exactly seven p.m., there was total silence as both sides held their fire. This ceasefire agreement was broken several times by both sides until the twenty-fourth of October when the United Nations Security Council reached a resolution to impose a total ceasefire.

The next day, all of the soldiers felt that the war was over. Israel lost close to 3,000 soldiers, which was a huge number in proportion to the population of three million. There was no doubt that Ariel Sharon was the Israeli hero of this war who caused the tremendous change from near-defeat to victory. The Egyptians totally surprised the Israeli army by launching this war, crossing the canal and advancing ten kilometers into the east side. This was a shock for the Israeli high command and it was well known that the Southern Command General Gorodish, who was the top commander of the Egyptian front, became confused and could not manage the war.

At the end of the *Yom Kippur* War, the Israeli forces were just 101 kilometers away from Cairo. Ariel Sharon was the leader who initiated the plan of crossing the canal and pushed the high command to approve it. In this war and especially during the crossing, I had the privilege to have seen him a few times. To the average soldier, he behaved like a private and ran around like a plain soldier. He was cool, courageous, and had vision. His morality and concern for his soldiers was above and beyond. He always encouraged the soldiers around him and raised their morale. He smiled everywhere he went and enquired about the situation. He never obliged anyone to salute him and talked to us as if he was a close friend.

If the Israeli government would have approved to continue the war for an extra day or two, the whole Egyptian army would have been destroyed and the Israelis would have conquered Cairo. The war was over and the situation was calming down. We met colleagues from civilian life and mandatory service and exchanged news about mutual friends. Common questions were, "Do you know so and so? Do you know what happened to him?" Many times the answer was, "I am afraid he was killed…. this one was missing…that one lost his leg… another one was in critical condition in the hospital." These were unpleasant moments that caused one's body to shudder. It was scary for me to think what happened to friends of mine whom I lived and worked with. I was eager to return home.

When Menachem received a list of all our battalion casualties and the hospitals where they were recovering, he assigned me and the deputy battalion commander, Captain Gershon, to go and visit each one of them. We got a list of over sixty casualties and divided them between Gershon and myself. My assignment was to visit the half of the casualties who were hospitalized in northern Israel, from Rambam Hospital in Haifa to Belinson Hospital in Petach Tikva. Captain Gershon was assigned the southern zone.

We received a short briefing instructing us how and what to tell our wounded friends about their colleagues in the battalion, such as who got killed, who is missing, and who got injured. We then went on our way to visit the wounded. We left the base on Saturday afternoon to Tel Aviv. When we passed the city of Ashkelon, which was Captain Gershon's hometown, he decided to leave our driver and me to visit his wife and son. He said to us, "I will complete my assignment with my civilian car."

The driver arrived in Tel Aviv at around eight o' clock in the evening, which was after the visiting hours of the hospitals. All I could do was to sit down near a telephone, and call the families of my friends who remained at the frontline, as requested since they knew that I was going to Tel Aviv. They gave me all the contact information of their parents, wives, and girlfriends and asked me to call them and pass on their regards. They were very happy to have this opportunity to let their significant others they were alive.

That night, I released the driver to return to his home until the next morning. I went to my home in Yahud in excitement. As I reached my house, the street where I lived was totally dark. Many children were outside and stared at my vehicle in curiosity while ignoring the person actually driving it. The commercial van was used for the war and was totally covered in sand and shrapnel. My brother, who was among the children, did not recognize me since I had a long beard and unkempt hair. I called out his name and after looking at me for a few seconds, he realized who I was.

He ran towards me and jumped into my arms. My parents were not at home and my brother did not answer when I asked him what was going on with my friends. He either did not know or did not want to tell me. I sent him over to my girlfriend who lived a block away to tell her that I have just arrived. I then entered the bathroom to take a shower, but did not shave my beard until the end of my service as a reservist. My girlfriend arrived and after we talked for a few minutes, we decided to hang out in Tel Aviv. Before leaving home, I asked my brother to help me in my assignment. I instructed him to prepare separate lists for the wounded in each hospital. I expected this to make my mission for the next morning easier.

My girl friend came with me to Tel Aviv and found a quiet city with dark streets. Almost no one was seen outside except for a few scattered soldiers in uniform who took advantage of their short vacation to watch movies. My girlfriend and I did the same and afterwards, we returned to Yahud. The morning after, I planned to pick up the driver and start my hospital visits. Before that, I decided to go to my old place of work in Yahud for just a few minutes to check how my co-workers were doing. The girls there welcomed me enthusiastically. They said, "Finally, the first man has returned to work." I answered them, to their dissapointment, "No. No. I am just here for a visit."

My first question was how my co-workers were doing. The girls looked at each other wondering if they should tell me or not. I repeated my question several times and judging from their nervous glances, I felt that something terrible had happened. At the end, they finally said to me, "Eliezer, the department controller, was killed during the first days of conflict on the northern front with Syria." Eliezer was a very tall tank commander who was a professional basketball player for the team of Petach Tikva.

I left the factory with unpleasant feelings. A whole day of work was in front of me and I did not know how and from where to begin my mission. I called the driver, who was named Moshe, to prepare himself and wait at the start of the highway to Haifa where I would pick him up. We started our visits with Rambam hospital in Haifa and then moved back towards the south.

I dedicated about five minutes to each injured soldier and reported to them the state of their unit, commanders, and soldiers. Some of them asked me about soldiers who were killed, but they did not know yet. I wondered whether to tell them or not. According to the briefing that we received on the base from the battalion commander, we had to encourage the wounded and raise their morale as much as possible. I could not tell them about the soldiers who were killed in action.

I did not know some of the wounded before visiting them and I found a few of them in very bad condition, such as missing limbs or with severe burns on the face or elsewhere on the body. A few seemed to be unconscious. Others had passed away on the same day that I arrived. I felt very sad and left each of them by saying "See you back in the battalion. Your colleagues are waiting for you and the battalion is getting reorganized." Judging from their condition, I doubted if they would ever get a chance to return to service.

On Sunday night I finished all of my hospital visits and called Captain Gershon to coordinate our return to the base in Sinai. He told me that he was not ready to return, since one more hospital in Be'er Sheba was left for him to visit. We scheduled to meet tomorrow morning in Ashkelon, go together to Be'er Sheba for the hospital visit, and then return to our battalion in Sinai. I passed the rest of that evening visiting relatives and friends and then went to bed in exhaustion. The next morning, I went to Ashkelon to pick up Captain Gershon and went to our last visit to the hospital in Be'er Sheba.

To our surprise and delight, we met General Sharon who came to visit some of his wounded soldiers. After we shook hands instead of saluting, he asked us personal questions and was very happy to hear that we had visited each of the wounded soldiers from our battalion.

We returned to our base in Sinai to report what we saw and heard on our mission. With this, I finished my service as a reservist and returned to civilian life. At the end of the war, I found myself a part of a sad nation.

CHAPTER 25

The *Yom Kippur* War, with its terrible bloodshed, was over. There wasn't one neighborhood throughout the whole country that did not lose loved ones. This was a severe tragedy for the people of Israel. Everywhere at home and at work, people talked and argued about the war. Accusations emerged against the military and political leaders that the Israeli army was not appropriately prepared for this war and the Egyptians had totally surprised us. The Israeli *Knesset* and the media demanded an establishment of an independent investigative committee to examine these allegations.

I returned to my work in the Aviation Industries after being absent for two months due to this war. The girls of my department treated me very nicely. During each break, we all talked about the war, about those who had worked in our plant who had either lost their lives or were injured. They were really missed. There was plenty of work to be done after the war, especially because the army lost a lot of equipment and vehicle parts. I worked in the field of chemical electroplating and became an experienced laborer.

We worked overtime every day and within six months, I was promoted twice. The plant's manager was Mr. Portugali. He was highly educated and experienced. He thoroughly supervised each department and every employee. He demanded maximum efficiency and accuracy at work.

One day, Mr. Portugali approached me with a special request. He was holding a recovered part of an anti-tank missile that was made in Russia and used by the Egyptian army during the war. He said to me, "I know this is not part of your usual work but please, do me a favor and try to disassemble this as carefully as you can, so that we can study how it was designed. I performed this task very patiently and dismantled the missile part as much as I possibly could. I transferred it to the boss and he was very satisfied of the accurate work I had done.

All of my work at the plant was completed through great will. I had a special insulated room for electroplating electronic cards and two female assistants helped me. Since my job was close to home, I was willing to work overtime every day and compensated appropriately.

I didn't have much time to sit down and have discussions with my parents during weekdays. I could had time on Shabbat itself. On one of these occasions, Dad said, "Listen, my son. The war is over. You have had a taste of military life. Now is the time for you to move to the United States to try your fortune. All of your friends from Afghanistan did not serve in any army and now some of them are successful businessmen in America. You see that how your life will be here in this country. No real peace is possible with enemies who are obsessed with destroying and killing. As it looks to me, you have no future in this place which is always in a state of siege. My friends in America want you to work for them, so you already have a place you can go. Listen to me and do not miss out on this opportunity."

I was deeply moved by this speech and told my father, "The truth is that you are totally correct, but please understand me too. We just finished the *Yom Kippur* War and I feel that this is my country. I do not want to switch the Jewish state with any other country. I also have a good profession here and I feel satisfied with the way things are. We will pray to G-d that no more conflict will erupt and everything will be alright." Dad was still not convinced.

It was already the beginning of 1974 and reservists from General Sharon's division had received summonses to return to their battalions. The summons that I received was for a forty day service. I requested my supervisor's assistance to try to delay the deployment date, by using the excuse that I was needed in my workplace which produced equipment for the military. Unfortunately, his intervention did not help.

I dressed up in my military uniform, picked up my duffel bag, and traveled to the Sinai Peninsula and then to the city of Ismaeliya. I continued further to the city of Faeid on the African side of Egypt where my unit was stationed. This city was the location of the headquarters for our division too. I was scheduled to meet Itzik Shakhed, General Sharon's administrative officer. He was advanced in age and had served for many years in the regular army. When I approached his office, I realized it was a military bus that had been converted into an office. Nearby, a tent had been erected for residency. Itzik said to me, "You will serve in this office for the personnel unit. Here are the keys and I want to see you here everyday." I went to the nearby tent that was to be my new residency and found there another three soldiers who served in the staff of the administrative officer.

During the day, we were quite busy and from time to time, we would attend officers' meetings on the bus. At nights, we had nothing special to do so we sat outside between the tent and the office where we would light a bonfire and sing Israeli songs. With time, we got impatient and asked each other, "What are we doing here in Africa? When will the Israeli army be finished here?" The Israeli media talked about the benefits of leaving the African side of Egypt and returning our forces to the previous border at the eastern bank of the Suez Canal. Official representatives of Israel and Egypt had many meetings in Sinai in the renowned 'Kilometer #101' for this purpose, but had yet to reach a settlement.

Occasionally, I had to take the administrative officer to the headquarters of General Sharon in my car. It was also in the city of Faeid. They had long meetings and I took advantage of this time to tour this beautiful city. Some buildings had been destroyed because of the war and others were occupied by Israeli soldiers. Outdoors, there were a lot of lawns and beautiful trees. Egyptian shepherds were seen with their flocks wandering from one place to the other. At every crossroad of this city, Israeli soldiers could be seen. I recognized some of these soldiers as past acquaintances and friends from my mandatory service and civilian life and we were very happy to see each other again.

During this reservist service, I received a weekend vacation once every two weeks to visit my family. I left the base on Friday morning and returned on Sunday night. It was a very long trip to Tel Aviv. The administrative officer, Shakhed, got along well with me. When we were off-duty, we used to sit together to talk and joke. I remember one day, two soldiers, who were on mandatory service, arrived at our base to serve as drivers. It made me happy since I thought that Shakhed would not use me anymore as his personal driver or errand boy. I talked with him and he agreed to release me from my driving duties.

Unfortunately, he changed his mind after a few days and he said to me, "From now on, you will be my personal driver for the remainder of your service." I had no choice but to comply since he was my commander. He explained to me that these new drivers were not familiar with the area and in one evening drive to the division's headquarters, they got lost due to the darkness.

One night that seemed to be as dark as the biblical Egyptian darkness, Shakhed called me to urgently drive him to General Sharon's tent. Within a short while, I succeeded to manage my way to the tent, even though I could easily lose my sense of direction due to the absolute darkness around us. During this drive, I was scared and it was a miracle that I even found the tent. Shakhed exited the vehicle and asked me to wait at the parking spot until the meeting was over. I complied and nervously waited for his return. I was very concerned with how I would make my way back.

After fifteen minutes which seemed to last for almost an eternity, Shakhed returned and ordered me to drive him back to the base. Darkness was all around us and I could not find my way back to the road. I then saw two barrels painted in white and I decided that this was a sign for the parking lot exit. I advanced and accidentally drove the car into the garbage dump of our battalion. Shakhed lost his temper and yelled at me, "You are no better than those other two young drivers. Look what you have done." I said, "Just give me one minute and I will get us out of here. I just lost my sense of direction because of this absolute darkness, in which it is impossible to see more than a few meters in front of you."

Fortunately, another vehicle arrived and entered the parking lot. This was a guide for me to take the right direction to get on the road which led to our base. This was one of the more unusual events that I experienced during that month.

Nevertheless, this unpleasant event with Shakhed did not ruin our positive relationship and he did not want to give me up as his driver. That is how my 45 day reserve service in Egypt elapsed. Close to the end of this service, we suddenly received an order to leave the African side of Egypt. This order applied to all military units of the Israeli army. Our unit had to print the detailed instructions for departure and we would then distribute it to all bases of the armored forces in this neighborhood. My unit was the last one to dismantle the equipment and leave Egypt.

At the end of May 1974, I again returned home to civilian life. On Friday evening before the beginning of Shabbat, my Dad and I were sitting on the porch overlooking orange orchards and hilly deserts in the far distance. Dad always used to sit on this porch and enjoy the view.

He suddenly became serious and gave me the following speech, "We have been living in the land of Israel for almost seven years in which you have spent four in military service and wars. In my view, this will be the pattern of your life here. You will always be called for reserve service in dangerous places. Your work in civilian life is not much better. Any amount that you earn, half goes to taxes of many different names. Here, you will not have a calm life or spiritual tranquility. Listen to me carefully my son. You know that I want the best for you. You are still young. Go and try your fortune in America where many of your friends went to succeed in business. No more military service and tensions. You have already completed your military service. Congratulations. You also served as a reservist during wartime and miraculously lived through it. Just think of it as a journey and if you don't like it, you can always come back."

This time, I really felt that I was in a critical dilemma. I did not know what to decide. On one side, we were in the Jewish state, the only land on the planet that belonged to us, and I had already gotten used to all of its difficult situations. I loved the life with my family and friends and I did not want to separate from them. On the other hand, I did not want to hurt Dad's feelings. Everyone considered him to be a very wise man who always thought a few steps ahead. New York was also the city where every discharged IDF soldier wanted to reach after the *Yom Kippur* War and fulfill the American dream. Crowds of them were waiting every day in huge lines at the American embassy in Tel Aviv to obtain a visa for the States. Consequently, the Israeli army decided to stop giving permits for discharged soldiers to travel to America.

These young men did all they could to get these permits with all kinds of arguments. At that time, I was busy with how to advance in my work in the Aviation Industries. I had a lot of satisfaction from this work, even though my salary had significantly shrunken after all the taxes. Dad frequently asked me how much my salary was and how much I had already saved in my bank account. My Dad said, "You have to get married and start a family and for that, you need money." Dad would repeat such questions and remarks very often. I always answered, "Don't worry. Everything will be okay."

I never told him that I had no savings in the bank. I was already twenty-four and trying very hard to save money, but the taxes as well as sharing expenses with my family left me with no savings for myself. The possibility to save money from work was exactly what Dad meant when he said that I would find good fortune in America. My father had old friends from Afghanistan who were members of Meir Shamash's family, and had succeeded in the American gem trade. They used to visit Israel often for vacation and had tried to convince Dad to let me join their business. They needed more employees and preferred to hire relatives or close friends. Whenever Dad came home from meeting them, he would take me aside for a personal conversation saying, "Michael, this is the chance of your life. You must go to New York. There is a Jewish saying that a change of location leads to a change of luck. Here in the land of Israel, you will not advance and you will only make the minimum needed to survive."

This issue kept me deep in thought. I consulted with friends and relatives and they universally encouraged me to take this big step. They often said, "If I was in your shoes and the same age, I would certainly go for it. What do you have to lose? At worst, you will just come back to Israel. This land will never run away." It was well known that every discharged soldier would like to leave the country even with no plans. Popular destinations were New York and Australia. The net pay for young men was low and due to nonstop conflict, every discharged soldier was called to serve in the reserves for over a month each year. This was enough to leave everything in Israel and look for a more promising future.

After the *Yom Kippur* war, the Israeli airline, El Al, had foreseen an upcoming mass immigration of youth to New York and purchased a new jumbo jet to accommodate this trend. All of this was shown on T.V. and received an unusual response by the American embassy and the Israeli army. It was almost impossible for a youth to get an exit permit or a visa to the United States. I knew that my father, relatives, and friends wanted me to go to the States and did not receive my objection well.

Nevertheless, in spite of the wars and hardships, I felt great in Israel. I loved my work as well as my colleagues and I felt content with my life and family. I used to hang out with friends and family in parks and celebrate all Jewish holidays together. I would never be able to find a life like this anywhere else in the world. This was the kind of life that I was taught to appreciate back in Afghanistan and a tradition started by our forefathers. I felt that this would be very difficult for me to leave behind.

The village of Yahud was small and everyone knew each other. My family was popular there. My brother Amnon was the manager of the local soccer team named Ha'Poel Yahud and he was very successful. My family accompanied the team to all of its games to cheer them on. Yahud started as a beautiful, quiet village, and we watched it advance from day to day and from year to year until it expanded into a town. The mayor, Mr. Chatoka, was a friendly person who made efforts to develop the town dramatically and bring it to the attention of the media. He indeed succeeded in his endeavors.

I was enjoying the summer of 1974. I loved the life in my small town and enjoyed hanging out with my friends. We used to travel all over the country from Metulla in the north to the port of Eilat in the south, and we toured the holy city of Jerusalem. I also traveled with my girlfriend on my scooter all over the country.

This enjoyable life came to a halt when my parents were invited to comfort the Shamash family for the passing of Meir's mother, which happened in New York. Her coffin was brought to the holy land accompanied with her entire family. It was on a Thursday evening and Dad returned home in anger.

The next day, Dad went to work as usual and in the afternoon, before the Shabbat eve, he sat down on his porch overlooking the beautiful landscape. He then summoned me and summarized the conversation he had with Meir Shamash the previous night. Meir had asked him, "Well, what is going on with your son Michael? When is he coming to New York?" Dad answered, "I guess soon. He hasn't yet decided since he had just been discharged from the reserve service and is still stressed out. I hope that during this summer he will come to a final decision."

He turned to me and said, "Listen Michael. This is the last time I will bother you in this regard. My friends are offering you a job in New York and it is unwise to reject it. Let's make a deal between ourselves." I curiously asked him what this deal was. "You will go on a trip for one month on my expense. If you like New York, stay there. Otherwise, come back. I will also pay for the return trip." This 'deal' made me feel like I couldn't disappoint my father and I gave my consent.

On Sunday morning, I went to a travel agency"Miriam" in the city of Petach Tikva. The agent there told me that in order to start the process of acquiring a ticket, I would have to submit a passport, an official wedding invitation from my family in America, and a one hundred dollar processing fee for the office. I immediately gave her the money so she would start processing. I then went home and made several calls to America in order to get a wedding invitation. After one week, I received an official invitation to a wedding hosted by a renowned jeweler which would take place on June twenty-seventh, 1974 in New York.

I happily returned to the travel agent and submitted my passport and invitation. The agent then told me that with this she could easily achieve a visa for me to the States. However, two additional documents were required: a release letter from my place of work and an exit permit from the army.

The first letter was easy to acquire. I had an interview with the security officer of my plant and he approved me for a month of vacation with no pay. The other letter was a lot harder to achieve. The army placed obstacles for young men to leave the country, especially to the States. I traveled to the Jaffa office that issued army permits. I already had the plane ticket with me and the date of departure was June twenty-fourth, 1974. Plenty of young men were standing in line nervously.

My turn arrived and the military officer began interrogating me. He asked me, "Why do you plan to go to America?" I answered that I was invited to a family wedding and I showed him my invitation, tickets, passport, and visa issued from the American embassy. The officer examined the invitation and asked for the identities of the couple to be wed. I informed him that the groom was my cousin and the ceremony would be the following week. The officer again examined the English invitation, stared at the date of the wedding, and said, "You are right. How long is the duration of your visit?" I told him I expected to be in America for only one month. He immediately gave me an exit permit that would allow me to leave the country.

All of the young men around me could not believe how quickly my request was approved. They were genuinely envious. When I returned home and told my parents that everything was set for my departure, my Dad was happier than I had ever seen him. "I told you that everything will turn out for the best. Do you believe me now?" Exactly seven years elapsed from the time my family immigrated to Israel in June 1967.

Rumors of my departure spread among my family and friends and they came to bid me farewell. The most difficult experience of all was separating from my mother. On the last Thursday of June 1974, I headed to Ben Gurion airport with a little suitcase and four hundred dollars in my pocket. I took an El Al flight to New York with a one hour stopover in Paris. I wouldn't have made this move on my own, but I did it for my father, who wanted this from the time we arrived in Israel. This seemed to be my destiny, to have such extraordinary experiences in Israel and then leave.

After an eleven-hour flight we landed in Kennedy airport. I arrived in a new country that had a different language and people who had a totally different mentality. I had to start all over from the beginning, in case I didn't want to return to Israel. The border control went smoothly and my sister and her two daughters, Teroa and Vardit, welcomed me at the airport. I got in her car and we traveled to their home. They had arrived a year earlier and lived in a nice house.

They lived in Rego Park in Queens, New York where many Israelis and Jews reside. My brother-in-law Simon was on a business trip away from home and I spent my first Shabbat in America with my sister and nieces.

On Saturday morning, I went to a synagogue which was a ten minute walk from the house where I was staying. Many of the congregants there were of Afghan origin and all of Dad's friends from Kabul were among them, like *Mollah* Meir Shamash, Shmuel Aharon, Babajan Vardi, Meir Simantov, Abrashk, Maidi and more. All of them were successful in the precious gem and diamond business and passed the 'business school' of *Mollah* Meir Shamash who had immigrated to New York in the early fifties. The prayer service lasted for over three hours, much longer than it would in Israel. Sermons by the rabbi and announcements by the president extended this service.

I felt very comfortable in this synagogue because I remembered many of the people from Afghanistan, and because of the plush upholstered seats, on which I almost fall asleep.

Right after the prayer service, I met Meir Shamash in the lobby of the synagogue. He immediately said to me, "So you have finally arrived. Your father is a very close friend of mine and I guess that you now understand why he wanted you to come here so much. I want to see you in my office Sunday morning."

Meir called over his office manager, *Mollah* Gavriel Aharon, and told him, "This young man will start to work with us as of tomorrow." Gavriel instantly responded, "But tomorrow is the famous wedding!" Meir then changed my first work date to Monday. I then attended a nice *Kiddush* which the synagogue had hosted on a regular basis after the Shabbat prayer to spoil the congregants.

At the end of Shabbat, my sister said to me, "Tomorrow we have a wedding to attend. You have also received an official invitation. Unlike in the land of Israel, every man who attends a wedding should wear a suit and tie. Let's go tomorrow morning to the famous department store, Alexander's, and buy two suits for you. You can also wear these suits on Shabbat and holidays." The store was right across the road from my sister's house and I also bought two ties and a shirt for Shabbat. On Sunday evening, I went to the wedding eager to know what they are like in New York.

Since the groom came from a family of jewelers, many of the participants were Jews of Israeli, Afghan, and Buchara origin who were in the diamond business of New York. What impressed me most was the order and style of this wedding. They had an orchestra of ten musicians and delicious food in large quantities. I was happy to meet old friends of mine there from Afghanistan who were from the families of Abraham Nissani, Maidi Aharonoff, and Simantov from Kabul. All of them were acquaintances of my father.

In the following evenings, they invited me along to a trip to the city. I was amazed at the sight of the Manhattan skyscrapers and the lights of Broadway. This was a very big city where everyone seemed to be in a rush. The subway system was the preferred method of transportation. Without it, New York would be paralyzed. Hundreds of thousands of people commuted day and night. My visit took place during the World Cup soccer competitions, and most of the world was occupied with it. To my surprise, United States wasn't. This type of sport was not popular here while the Americans were enthusiastic over other sports like basketball, baseball, and American football. Baseball and football were unknown in Israel and hardly ever played in Europe.

Early Monday morning, I took the subway from Queens to Rockefeller Center in Manhattan. Nearby, at the corner of 47th street and 5th Avenue, was the office of my new boss, Meir Shamash. His company name was Ambuy. This was a huge office that occupied half of the floor. It was furnished in a modern style and they had a staff of six, and my boss. Meir welcomed me warmly and said that he had just spoken on the phone with my Father. He added, "We agreed that you will study business of precious gems here and later on, you will pass the knowledge over to your brothers who will join the business too. Meanwhile, the money that you will earn here will help your parents in Israel. This is a good business with an excellent future."

He approached a safe box and took out a package of rubies and a package of sapphires and spilled their contents onto a desk. He then said to me, "Sit down and sort and separate between gems with different colors and different qualities. From now on, this will be your desk. When my *Akesh* Yehuda returns from his business trip in the Far East, we will sit down together and discuss your salary." I sat down enthusiastically and sorted and studied the gems.

The company's secretary, an elderly woman with a lot of experience named Edith approached me with two forms. She said that I had to fill them out in order to achieve a green card from the immigration service, as well as a social security card. All of this was needed so I could start to work legally in the United States. She said "In case you have any difficulties with it, I will help you out." It took me about two hours to complete these.

I only needed a letter from my new place of work that they

wanted to hire me, and that I had the appropriate qualifications for the job offered. The secretary gladly typed a letter and Meir Shamash signed it. Even though I knew that it would take a few weeks until my applications would be approved, I was confident that I had the job.

Occasionally, Meir came to my desk to check on my work and he always encouraged me that I was advancing rapidly. Sometimes, he invited me to lounge in a restaurant downstairs where most of the stores sold jewelry and diamonds. This was the famous 47th street which is known all over the world. He once said that he wanted me to get some experience in sales. "You will go and knock on the doors of the jewelry storefronts, and get orders for our company."

Meir did not speak Hebrew, only English and Afghan, so we conversed in Afghan. He had a strong personality and sophisticated business acumen. Of course, he also possessed a lot of good fortune since he had become a very wealthy man from his business. He had actually started exporting gems while he living in Afghanistan. I did not like the sales part of my job. Business was slow and therefore it was very difficult to obtain orders. However, I left business cards at every place I visited with the hope that in the future, these businesses would need us.

Nevertheless, our work was hectic since many salesmen from Japan and Thailand came to offer us their polished gems. Most of these salesmen were of Afghan or Persian origin. We checked each offer very carefully and did not deny any salesman in advance. If it was a good bargain, Meir or Yehuda Shamash approved the purchase and placed the gems in the safe.

One of the office secretaries was a relative of the bosses, Meir and Yehuda, and an Israeli. She liked to speak to me in Hebrew and really wanted me to advance in my new career. She approached the bosses and said to them, "I realized Michael is a good asset for the office and we have to help him improve his English. What do you think of sending him to an English course in the evening?" The bosses immediately gave their consent to pay the tuition fee.

The next day, she accompanied me to the Rohd's School office on 5th avenue which was close to our office. I registered for an English course that I had to attend three evenings a week. I also obtained a library card so that I could get required books for my studies and do my homework there.

Not much time had passed until I adjusted to my new life. It was totally different from the life I experienced in Tel Aviv or in Kabul. The mentality of Americans was different, but they were more generous, cultured, and efficient. In Israel, my family and friends thought that I wouldn't survive in America, and I believed the same. Much to my surprise, this was totally not the case. I got used to my new lifestyle from the very start. The bureaucracy in America is much less confusing than it was in the previous countries I lived in. Waiting in lines took less time as well since there was better organization.

I made new friends, especially with people who grew up in Afghanistan. Every weekend they knocked on my door and invited me to hang out in the city with them. Usually, we went to see a movie on Broadway or to an Israeli club in the famous neighborhood known as 'The Village'. Sometimes, we just went out to wander the streets of Manhattan or sit in a restaurant for dinner.

One Sunday I was home when suddenly I heard a knock at my door. I was shocked and could not believe who it was. A former Israeli boss of mine, Mordechai Pozner, stood before me. It was during the years of 1968 and 1969 before my military service when I worked in his Relif plant. We liked each other very much. I asked him in astonishment, "How did you find me here?" He answered, "As I was planning a trip to America, I decided that I had to see you. I called your parents to get your address and that's how I found you." We sat together for an enjoyable talk and then he asked me to join him with his family to a show in Radio City Music Hall. I was happy to receive this invitation, and in the afternoon, we took the subway to Manhattan. It was a live show followed by a beautiful movie starring Julie Andrews.

We returned to Rego Park later and went to a steak house together. It was the only kosher restaurant in the neighborhood at the time. Their steak was the best I ever had in my life. I had successfully adapted to American society in every way.

New York is like a huge factory that is operational 24 hours a day and people hurry nonstop everywhere they go. I had happily become part of this system. I always liked the American mentality and generosity since I was a child in Afghanistan. We lived near the American embassy in the city of Kabul and I used to take part in all activities it offered for the local public. Since then, I had the strong impression that the Americans were good-hearted and wanted to help the whole world to be a better place to live in.

After a few months in New York, I had familiarized myself with all the main streets and could travel by myself anywhere I wanted. The system of highways, overpasses, tunnels, and bridges were amazing to see. They efficiently connected different parts of the city as well as to other cities and to the nearby states. At that time, it seemed to me like a great wonder since Israel did not yet have such a system.

Fortunately, I became part of New York's large united community of Afghan Jews. This gave me a content feeling that I wasn't alone, but rather part of a tightly-knit group. Most of them lived in my neighborhood, and I met them locally or at my job, where many of them visited for business or consultations.

At my job I learned new things everyday regarding gems and the business of dealing with them. I listened very carefully to my bosses' instructions and I had to be very accurate with my work. I did my best to advance in my field, but was aware that many people who were in a position like mine had given up within a few years. I was in my position for just a few months, and was curious to know how it would develop. I waited for my green card to arrive for months but didn't hear anything about it.

Almost a year had passed since my arrival in New York in 1974 and I finally received a letter from the immigration service. I opened the envelope in excitement and I was shocked. Their letter stated that I had come to the States for a trip of up to three months, and since I violated the conditions of my visa, they were not willing to prolong my stay, and I had to go back to Israel on a specific date.

I took this terrible letter to my workplace but Meir was absent that day. I fearfully visited the office of his brother Yehuda and showed him the letter. He calmed me down and said, "Don't worry. I will take care of it right away. In the meantime, you just continue with your work and don't get into trouble with the police." He immediately called an expert immigration lawyer. The lawyer listened to my story and then advised Yehuda to write a letter, stating that I was needed due to my professional expertise in gems and knowledge of the Afghan language which was essential in his business. This letter would be submitted by the lawyer to the immigration bureau. The lawyer had a few negotiations with the officials which reached a successful conclusion. I was approved to remain in the States but was not yet given the desired green card.

I had been successfully working for over a year. During this time, I really missed my parents and siblings who were still in Israel. My Father's friends in New York occasionally invited me to their homes. I felt like family around them. My manager, Gavriel, always invited me to his home on holidays and pressured me to marry during these occasions. He said to me, "We have plenty of young women in our Jewish Afghan community here and if you give your consent, we can make an arrangement for you." I thanked him and said, "I am not worried. My time will come soon."

One day, my sister who lived in New York, matched me with a young woman who belonged to a rich family in the community. We scheduled a date and she arrived with a luxurious car which impressed me. On our second meeting, which was on a Saturday evening, she invited her brother and me to a show featuring a renowned singer, Diana Warwick, in Manhattan. We met for a few more dates, but I wasn't sure if she was right for me.

This continued until one day, her uncle approached my boss *Mollah* Meir and said to him, "We have to urge this couple to set an engagement date. My boss, who was a very smart man, pretended that he did not know we were dating. "Michael didn't tell me anything about her. I will talk with him to see what is going on, and then we'll see where to go from there." The next day, Meir called me to his room and went straight to the point, "Michael. Do you love the girl you are dating now? Answer clearly yes or no, do not play games." I answered, "Sir, I do not love her yet." Meir got serious and said, "End the relationship with her immediately and don't waste her time. She is from an important family and is serious about dating. You are not ready to make a decision yet, so stop seeing her." I complied with his demand since I sincerely felt he was right.

In Israel, my brother Ishai got engaged. I received an invitation for his wedding but could not go since I would not have been permitted to return to America, since I didn't have my green card yet. It was the summer of 1976 and I was living in an attic with another young Afghan guy named Avi Kolangi. Each of us paid $120 a month for the apartment to my bosses. The rooms were neglected and had a lot of dirt and cockroaches. No matter how much we cleaned and exterminated, it didn't help. Avi and I decided to look for a one-bedroom apartment in better condition. Fortunately, we found a very nice apartment in a luxurious building in Park City, Queens.

The monthly rent was $320, which was quite affordable. Avi was a hard laborer who worked seven days a week until ten p.m. every night. I never actually saw him except on holidays, and sometimes on Saturdays. He was a quiet and polite young man who was easy to get along with. He performed the cleaning in the apartment while I cooked the food I bought, and we both ate it. The new apartment had a large balcony with a view of the street.

Soon after we moved in, our friends started to visit us on Friday and Saturday evenings to sit on the balcony. We served snacks and drinks and listened to Israeli music. It became so popular that our friends would bring their own friends and the amount of visitors grew week after week.

I received several offers for dates but I declined them. I did not feel like dating until I received my green card. I expected to receive it soon and I planned to visit my family in Israel. I thought to myself that perhaps I would find my soul mate there. Since I left my mother sent requests for me to return, telling that she had found a good match for me.

My plan to visit Israel was put on hold since I met a young Israeli woman named Dahlia who also worked for a jewelry company in Manhattan. She also had originally come to the United States to attend her sister's wedding, who resided here. On this trip, she received an offer to work for a jewelry company which she couldn't refuse. Almost every week, I met her at the post office where the two of us picked up our mail. She was living at her sister's house in Manhattan and was looking for friends.

In July of 1976, I met her by chance on the subway and she told me that August fifth will be her birthday. I told her that every Saturday night friends of mine meet at my apartment and I wouldn't mind throwing a birthday party for her. She liked the idea and agreed to it. I told her that I was going to Florida for vacation and that I would get in touch with her upon my return.

Most of the jewelry businesses closed for summer vacation for the first two weeks of July. I was entitled to two weeks of vacation from work and decided to spend the time in Florida. My sister's brother-in-law Rafi owned a few houses there and was visiting New York. I spoke with him about my plan to join him on his way back to Florida and spend my vacation there. He said to me, "You are more than welcome. You will enjoy crossing the whole United States from north to south to see the beautiful country in my car."

At the time, America was celebrating two hundred years of independence and there was an atmosphere of euphoria all over the country. This was a most appropriate time for a road trip. We loaded the car with food and took the I-95 highway down to Florida. Since it was a long journey, we took turns at the wheel. Rafi offered to drive slowly so we could see all the beautiful places on our way during this beautiful season. He said, "Don't worry. We have plenty of time, rushing isn't necessary."

We slept in a motel in North Carolina on the first night of our trip. The next day, we reached Cape Kennedy where we went on an organized tour. On the second night before midnight, we reached Rafi's house which was in the center of beautiful Palm Beach. Rafi had made a nice profit from selling and buying real estate and also tried to get into the gem business. He often came to New York to buy stones and sell them in Florida. Several times, he bought the gems from our company and I helped him with his deals. We had become good friends.

We spent together ten days thoroughly touring the whole state of Florida. We went to Miami, Ft. Lauderdale, Disney World in Orlando, and more. We also went on a boat ride at night. After many pleasant adventures, I took a flight from Palm Beach back to New York, while Rafi stayed at his home in Florida. I returned to work with a deep tan after being in the Sunshine State.

This visit made me realize the difference between New York and other states of America. People in New York are from all over the world and original Americans are hardly seen here. This is a city for hard work where people are obsessed with making as much money as possible in the shortest time. Everyone was trying to win his fortune and I was among them. Even New York's weather was unique. It's cloudy most of the time, and during the summer, it rains along with an oppressive heat.

August came and I met Dahlia again on the subway and I remembered the birthday party that I promised her. We scheduled to meet at my home on the coming Saturday evening and invite our friends. She arrived with her sister Molly before the other guests to help my roommate Avi and I prepare the party. Even though we didn't exert much effort, the party was very nice. All of our Israeli and American friends arrived and we ate, listened to music, and sang until the late hours. Dahlia thanked me and promised to come back tomorrow morning to help clean the mess.

The next morning, a phone call woke me up. Dahlia was on the line and I said to her, "It's still early. Why aren't you sleeping?" She answered, "I felt responsible to clean up the mess we left yesterday, so I woke up early this morning." I said to her, "Don't bother, I can take of the mess on my own." She insisted on coming anyway. Half an hour later, she knocked at my door. I got out of bed and let her in. I felt that matters between us had become serious. I thought to myself that maybe she was my soul mate. However, I had other prospects from Israel and from the local area which left me confused.

Objectively, I was alone in this foreign country and she was in the same situation. I decided to wait and see what the future had in store. I didn't really like going out much. When I went out to a movie or some other event, it was to accompany friends. I liked to sit in my corner of the balcony to read books or listen to the Israeli news on the radio. If we were a good match and lived together, it would be much nicer to be at home.

One Saturday evening, some of my friends and I took a few cars to hang out in Manhattan. On the way we listened to the radio about a successful Israeli rescue operation in Uganda. We immediately turned the cars around and returned to my apartment to sit on the balcony and listen to Israeli news. The local radio and TV news covered the operation but naturally, it wasn't as detailed as on the Israeli radio.

An Air France jet on route to Tel Aviv was hijacked by Palestinian terrorists and forced to land in Uganda. Most of the passengers were Israelis and the terrorists demanded the release of all Palestinian prisoners in Israeli jails. If their demands were not met, they would kill all of the Israeli passengers. Uganda was under the rule of the crazy dictator Idi Amin, who was an anti-Semite. This made the terrorists feel emboldened and safe at the international airport of Entebbe, the capital of Uganda. Israel sent a military cargo plane with a special commando unit led by Yoni Netanyahu to travel thousands of miles to Entebbe.

This unexpected operation caught the terrorists by surprise. They were all killed within a few moments. The hostages were released, brought to the cargo plane, and within minutes, left the airport for Israel. This amazing and courageous operation had stunned the whole world and inspired pride in every Israeli. Unfortunately, the only Israeli soldier who was killed in the operation was the commander leading the forces. This was named Operation Jonathan after him.

CHAPTER 26

My supervisor Yehuda planned to visit Israel. I asked him to visit my parents and invited him to my younger brother Ishai's wedding, which was scheduled to happen soon. My roommate was also going to Israel after being away from the country for many years. He had just recently received his green card which enabled him to leave the States. My green card was still pending. The immigration lawyer who was taking care of the process promised me that I would receive it soon.

In 1976, I decided to buy a car for myself. I hadn't saved a lot of money so I could only afford a used car. Every day, I checked the newspaper classifieds for second-hand cars for sale. Some of my friends helped me in my search. Finally, I found a Volkswagon Beetle from 1969 with a reasonable price of $450. One of my friends took me to Brooklyn to see the Volkswagon. When we arrived at the owner's home, a hippy drug-addict answered the door. I didn't understand his English and my friend helped me out.

The car seemed to be in good condition but the muffler made very loud noises. My friend took advantage of this defect and offered the hippy only $350. He said to him, "If you agree to this amount, we will pay you cash and finish the deal within one minute." That's what happened and I drove the car back to Queens while my friend followed me in his own car. When I returned home, he said to me, "Michael, this is a nice car and we got a great bargain." I used the Volkswagon with pleasure for about a year, until one day, after a long drive, smoke emitted from the engine. I immediately drove to a garage and the mechanic asked for a fee that was more than the cost of the car. I left the car for him as a present.

In 1976, I received a few match proposals from friends along with pressure from my family to get married. I wasn't sure if any of the proposals were suitable for me and I got confused. I had broken up with Dahlia but we maintained a positive relationship. One day, she came to me and said, "Mike, I have grown attached to you and cannot give you up!" I didn't know what to answer her, but after my experiences with other girls, I had come to the conclusion that Dahlia was not bad for me at all. We decided to resume our relationship, but with no commitments.

I finally received a letter from the immigration bureau requesting that my lawyer contact them to order schedule an interview to receive my green card. The significance of this was that I was able to visit Israel, after three years of not being able, and this was a cause for celebration. I planned to go for the upcoming *Pesach* which was the most appropriate time to visit my family. I consulted with Yehuda who had just returned from Israel where he attended my brother Ishai's wedding. He also notified me that my roommate Avi got engaged in Israel to a young woman who I knew very well. Her name was Rivka and she was from the Yazdi family from Afghanistan. This news was the biggest surprise of the year for me. I called the couple and wished each of them *Mazal Tov*.

My roommate Avi remained in Israel to organize his upcoming wedding with his soul mate. I stayed at the apartment by myself but on weekends, everything was as usual. Friends visited me on Saturday and Sunday evenings and enjoyed my cooking. The food they liked the most was an Afghan cuisine called *ash palow,* which contained rice and carrots.

A cousin of my bosses joined our group and gathered with us on weekends. He lived in Italy for many years and one day, moved to America. He offered me to sell me his car which was a German-made Opal. This time, I had really bad luck. It was in the garage for repairs most of the time and the expenses were tremendously high.

One day, I met a new mechanic who had an Arabic accent. I

asked him where he was originally from, and his instant response was, "Are you Israeli?" I confirmed that I was and he immediately went nuts. He ranted, "You Israelis with your Jewish lobbyist abroad rule the whole world. You robbed our lands by force." He then grabbed me by the jacket and pulled me over to him. He said, "How would you feel if I took your jacket by force and ran away?" I instantly seized his arms and pulled them off of myself. I then calmly explained to him, "First of all, you aren't strong enough to take my jacket, compared to me. Second, neither you nor I declared wars between our people. It is the leaders who we cannot control. I doubt if it's even their own decision, since they are compelled by the superpowers: the U.S.A. and the Soviet Union. What do you want from me personally? Is this all my fault?"

After we exchanged verbal clashes, he admitted his behavior was wrong and politely introduced himself. He then said, "I am from Egypt and took part in that horrible 1973 war. Miraculously, I survived." I responded, "Oh really? That is interesting. I was involved in the war too and stayed in Egypt for a few months. I was stationed in the Egyptian cities of Fayed and Ismaeliya." I noticed that he had totally calmed down and then I said to him, "At least maybe we could be friends now." He answered, "Sure. Why not?"

The mechanic checked the car and said to me, "I do not recommend that you invest more in this car. If you want, I can buy it from you for $350 and use it for spare parts. I am even willing to accompany you to a car dealership to help you find a better car." I gave my consent and we went together to a nearby car dealership where I bought a 1969 American Plymouth for $750. With time, I realized this was a very good deal.

I received a letter from Israel telling me that my grandfather Yehuda had passed away on *Hanukkah*. This bad news saddened me. He was always an optimistic person and had a *Ruach HaKodesh.* All of his advice was always good. Anyone who followed it succeeded and anyone who disregarded it failed. He was always ready with blessings. I felt very sorry that in my upcoming visit to Israel I would not see him.

A short while later, I received more bad news. One afternoon, I was sitting in my boss Meir's office when the telephone rang. On the other end of the line, a scary voice was heard. Meir just nodded his head and hung up. He then told me that his father in law, Mullah Avraham Shalom Shamash, had just passed away in New York and his funeral was being prepared. He was once the head of the Jewish community in Kabul and this bad news shocked Afghan Jews throughout the world.

I saw him awhile ago in our office building in the elevator. I had entered quickly and greeted him with a nod of my head. He then rebuked me saying, "Hey *Farzand*! You greet people with your mouth with the Hebrew word shalom, which means peace for the whole world. Say it out loud so I can hear it." Afterwards, he asked how I was doing here and how my parents were doing in Israel. He was beloved by the whole Afghan Jewish community and anyone who met him. He was a soft-spoken man and therefore, he was also called *Mollah* Shalom.

This was the end of 1976 and I was twenty-seven years old. Pressure to marry was on me from all around since it was deemed that the appropriate time had arrived. So far, the best match for me in my view was Dahlia. Even my boss who had strict views was also impressed by her. She originated from a family that was well-known in the Jewish community of New York and Tel Aviv.

Her uncle, Dr. Chadi, had a home near my boss Yehuda in Forest Hills, Queens. Another uncle named Nathan had a translation and public relations company on the famous Alemby Street in Tel Aviv. Originally their family came from the city of Kashan, in Iran, but Dahlia was born in Israel. Her parents lived in Kfar Saba, a town north of Tel Aviv. They were well-known for their charity work and her father was also active in public relations. I thought to myself that I first had to receive a green card, then visit Israel, and only afterwards deal with my marital status. Meanwhile, Dahlia wouldn't leave me alone, even though she was just an option in my eyes.

At work, it was quiet since the gem business had slowed. However, I had so far succeeded to acquire some customers from big companies, who placed their full trust in me in business matters. Almost every day, I was busy with some of them to sell stones. One day, I overheard an argument between my bosses Meir and his younger brother Yehuda. I had been sitting near their office while the door was open. They had a branch office in Thailand which had been busy at the time. They wanted to have a manager there who they trusted and had good qualifications. I heard Yehuda saying that he wanted me to go to Thailand. Meir agreed with him that I was suitable for the job, but he objected to send me off due to other reasons. Meir said to Yehuda, "Michael is the son of one of my close friends and I cannot put him at risk by sending him to a country like Thailand. New York is quite a safe place for him while Thailand is a totally foreign country with no trace of Judaism. If, God forbid, something happens to him there, I would have nothing to tell his father." That is how I remained at my current job in New York until I received my green card.

I received a certified letter that said my green card was waiting at the immigration bureau. I was so happy and excited that I had forgotten to call my job to tell them that I'd be late. I immediately dressed myself in a formal green suit and went to the immigration bureau to receive my green card. I then went to work and my boss, who noticed me arriving late in formal attire, stared at me and asked, "What's going on? Are you getting married?" I answered, "No. Just look how quickly I received my green card." Meir burst into laughter which I had never seen him do before and said, "You dressed yourself in a green suit just for that? I cannot believe it. Anyway, I wish you *Mazal tov*."

I asked him for a vacation since I wanted to go to Israel. He answered, "Not now but very soon. The holiday of *Pesach* is coming and you will get a whole month of vacation which you deserve." From that moment, all of my thoughts were on Israel. I missed my family and the country very much. I must admit that within the three years here, I got used to the mentality and the lifestyle.

During the last days before my departure, Dahlia visited me constantly. She reiterated, "How could you leave me?" On the other hand, she often stated that she was jealous of my trip. I told her that I was waiting to go back to Israel for a long time, but could not until I had my green card. I told her, "Don't be jealous. Come with me." She nodded but could not come.

Dahlia called her family in Israel to tell them for the first time that she had a boyfriend of Afghan origin. Their response was "Keep this friendship since Jewish Afghan men are known to be very good people." She told me that her parents and three sisters in Israel wanted to see me. I said that I was ready to do this, but made no commitments to her. She said that she would miss me and I told her the same.

A day before leaving, my boss called me to his room and asked me what was going on with Dahlia. He wanted to know if she was still my girlfriend. When I said that we were still seeing each other every day, he suggested to me that it was worth marrying. He added, "She is from a very good family." I then said to him, "Upon returning safely from Israel, I will seriously consider it."

In my last week in the States, I went shopping almost every afternoon to buy gifts for my family. I mainly bought clothing and electronics which were very expensive in Israel at the time. I bought a round trip ticket with British Airways with one stop in London. At the travel agency, I met a few more Israelis who had also bought tickets for that flight to Israel.

On the day of departure, I dressed in the same green suit I wore to pick up my green card and took two suitcases with me. Dahlia accompanied me to the airport and stayed with me there until the last second. I looked into her eyes and saw how difficult this separation was for her. I could not help her. This was part of life.

CHAPTER 27

On the flight, I was overwhelmed with emotion. I felt that this time was one of the best moments of my life. I was aware that this is a subjective feeling, but it was still real for me. A state of tranquility had befallen me. I was on my way back to the holy land where the majority of my family was residing. I imagined over and over again the upcoming reunion with my parents and siblings. I first counted the hours, and then the minutes until the final landing.

I arrived at Ben Gurion airport near Tel Aviv the next afternoon. After I passed customs and the border control, I met my family who were waiting for me at the arrival lobby. Half of my close family and relatives were there while the other half were waiting for me at my parents' home. After an emotional reunion, I entered my brother Ishai's car and sat between my parents.

A flood of questions was released from my parents once we were settled in the car. "We want to know how long you're staying." I told them that I would be with them for one month. Then Mom said, "We heard that you have a serious girlfriend in America. When will the wedding be?" I immediately answered, "If everything goes smooth, then the wedding will happen this summer." They asked where, and I answered that it would happen in America. Then they asked me, "Won't we get to attend your wedding?" I answered that they will certainly be invited if everything goes well.

My brother Ishai, who drove the car while listening to our conversation, interrupted saying, "You'll have to charter a whole plane for the entire family to come. I promise you that all of us will make it." He then burst into laughter. I hardly noticed how quickly the time flew and we had arrived home.

My parents' apartment was very crowded with little room to move. All of my brothers, sisters, uncles, and cousins had congregated in the apartment to welcome me. The children of the family had grown while I was away and I could hardly recognize them. My little brother Avi had gotten much taller and his face had changed. Everyone wanted to know how America was and how was I doing. As usual, Mom had prepared *ash palow* and served the guests. Cheers of joy were heard all over our home and everyone took turns greeting me and inviting me to their homes.

The guests slowly left and I remained with my parents. Dad was very tired and said that we would talk tomorrow since he had to go to sleep. I did not have much time to talk with him and ask him how his life and work was. The crowded party prevented such discussions. I urged Mom to also go to sleep since she looked exhausted after all the preparations she had done to welcome me. She insisted that she was not as tired as I thought she was and said to me, "I'm used to all this and don't forget that we don't have such happy occasions everyday."

I went to bed exhausted but could not sleep because of jetlag. Only at daybreak did I fall asleep and I woke up just a few hours later. Everything was quiet at home since only Mom and I were there. She was busy cleaning the house and preparing food for the upcoming *Pesach*. When she saw me awake, she came and sat next to me. This was an occasion for her to talk to me one on one, with no interruption. Her first question was my girlfriend's name and I told her. A sequence of questions followed: What was her origin? Who are her parents? I told her that Dahlia was born in Israel but her parents came from the city of Kashan in Iran.

She said to me, "I know two girls who would be right for you and I want you to see them." I declined, explaining to her that Dahlia was my soul mate and we are going to visit her parents who lived in the city of Ramat Gan. There was a sudden silence and Mom said no more on this subject. It was a bit difficult to explain my feelings, but when Mom offered her matchmaking assistance, on the spot I came up with the decision that Dahlia was my soul mate indeed. Maybe the reason was that after a close relationship, the sudden physical distance between us caused me to miss her greatly. From this I understood that she was my soul mate and I had no interest in anyone else.

I went out to the porch where Dad liked to sit on Friday afternoons. It was a wonderful morning with clean air. I looked at the orchards and smelled the sweet fragrance of the oranges that was pervasive at this time of the year. It was spring, everything was blooming and it was a few days before the *Pesach* holiday. I went back inside and saw Mom still cleaning for the holiday. I told her, "I cannot see you working so hard. I want to help you." She refused and said, "*Pasar Jan*, almost everything is ready for the holiday. We have done all the shopping and cleaning and only a little bit of work is left. You came here to relax and not to work. Anyhow, thank you for the nice gesture." I continued, "Maybe I can rent a car and we will go for a trip all over the country. I miss Jerusalem and I would like to see it again." Mom acted like she didn't hear me and said, "We have not had a *Pesach* with you since we left Afghanistan. For over three years, you were in the army and celebrated *Pesach* in your military bases. Afterwards, you were in America for another three years. This year, all of us must celebrate together. I have already invited all of your brothers and sisters and this holiday will be a double celebration." I listened carefully to what Mom said and realized that she was right as usual. She was the unifying power of our family and always wanted to see all of her children around her on special occasions.

That *Pesach* night was the best one in my entire life. We had the celebratory dinner and recited the *Haggadah* exactly as we did back in our hometown of Kabul and in accordance to the customs of our forefathers. The recital was done in the unique and beautiful melody. The next day, we went to the morning holiday prayers in the central synagogue of the Jewish Turk community. Many members of the congregation knew us very well since we used to pray there very often.

Rabbi Gavriel Bar Simantov, who also served the synagogue as its cantor, came to me after the service and said, "Hello Michael. Welcome back. How was it in America? Did you make a lot of money there? Did you save anything?" At that moment, I was surprised with his questions regarding money. What did that have to do with the synagogue? The rabbi noticed my surprise and immediately gave an explanation to his questions, "We urgently need to expand our synagogue since we are getting more members and as you understand, this needs money. If you could collect this money for us in the States, it would be a big *mitzvah.*" I instantly agreed and said, "I will do the most to help the congregation with this endeavor. I have wealthy friends in New York who could help and I assume that they would want to."

On *chol ha'moed*, my brother Ishai offered me his car to use for my personal enjoyment. Before I accepted, my brother-in-law Levi also offered to loan his car to me. I decided to accept both offers, one after the other. I then recalled that I had promised Dahlia to call her parents as soon as I arrived in Israel. I only had her sister Rachel's telephone number. She picked up the phone and when I introduced myself she said, "We heard a lot about you from Dahlia. When are you coming so we can meet you?" I answered, "I can come over whenever is good for you." Then she said, "We are expecting you any day and it would be great if you came over for dinner tomorrow night." I told her that meals were not necessary but I would enjoy some coffee. I also explained to her that during the *Pesach* holiday, my parents would only eat meals at home. She asked, "What? You're bringing your parents?" I answered, "Yes, and I would also like to invite my brother Ishai and his wife. Is this okay with you?" She agreed.

The next evening, all of us traveled to Pardes Katz, a suburb of Ramat Gan, to visit Dahlia's family. They lived in the city of Kfar Saba for many years and had recently moved. I knocked on their door and Dahlia's sister, Rachel, answered. I was amazed to see how similar she looked to Dahlia. She invited us in with a smile. We entered the living room and saw that the apartment was very small. Rachel's red-haired husband was sleeping on the sofa in the living room and greeted us when we arrived. He got up and invited us to sit down. I asked Rachel where her parents were. She told us that her father would return from work any minute and her mother was in the shower. In the meantime, another of Dahlia's sisters, named Judith came home with baskets of food. She said hello to us and went to the kitchen to put away the food.

When she came back to the living room, she loudly said, "Welcome to our home and we hope you are having a happy holiday." She looked at me and said, "Hi Michael. We heard a lot about you from Dahlia. How is life in America?" All I said that it was great. Dahlia's mother entered the living room, greeted each one of us, and sat in the corner. She was a quiet woman and did not have much to say. We were waiting for Dahlia's father to come home and chatted in the meantime.

We suddenly heard loud knocking at the front door and Dahlia's father appeared. He was dressed in work clothes with big boots like a firefighter's. All of us rose and welcomed him. Dahlia's father had a deep, strong voice with a real Persian accent. He was very friendly and talked about very interesting topics. It was a pleasure to sit and listen to him. A few minutes after his arrival, genuine Persian tea was served with dried fruit. We sat together for another hour and stood up to leave. Dahlia's whole family accompanied us outside and asked if we would see each other again.

On our way back home, we discussed the family we had just met. They were very decent people who immigrated to Israel from Iran in 1950. I wanted to hear my parents' opinion. Mom said that they looked like an honorable family, who were indeed poor but similar to our origin. From the stories we heard there, we had the impression that Dahlia's father had done a lot of favors for people, and in this regard, he was very similar to my parents. With time, I realized that this meeting had a very deep influence to the future of my marital life.

Dahlia was the fifth daughter in her family after four consecutive girls. She was born on August fifth, 1953. She was four years younger than me and was born in a little town called Kfar Saba northeast of Tel Aviv. A fifth consecutive daughter is considered extreme in the Jewish Persian community. Dahlia's father did not have on opportunity to visit his wife in the hospital when she was born. Because the father wasn't there, Dahlia's mother waited to name her. On the last day of her hospital stay, the nurse came to her and said, "Today, you are leaving the hospital and your newborn girl does not have a name yet. Don't you think this is ridiculous?" Dahlia's mother felt embarrassed and asked the nurse for her name. The nurse answered, "Dahlia." At the spot, Dahlia's mother said to the nurse, "This will be my daughter's name. I really like it."

The seven days of *Pesach* elapsed and I had one more week before I had to return to my work in New York. I had visited all the places and people that I had planned to visit in Israel. Before I left, Mom wanted me to also visit the grave of Grandpa Yehuda. She said to me, "He was a real *tzadik* (righteous man) who loved you very much. He is buried in Jerusalem in the cemetery at Har Hamenuchot." I called the *Chevra Kadisha* and asked them the exact location of the grave and then left to Jerusalem with a few of my friends. We first visited the old city of Jerusalem. We saw the Wailing Wall and other holy sites of the Jewish religion. We then went to the cemetery which was nearby but unfortunately, we couldn't find my grandfather's grave. I didn't know what to do and thought that perhaps Uncle Nafthali could help me out. I called his home and he gave me very specific instructions on how to reach the grave. He said, "From where you are, go down fifty meters and you will reach a fence and a water fountain. Go down a few steps and you will find the grave. It's definitely there."

My friends and I followed Nafthali's instructions but we still could not find the grave. We then decided to scatter and to examine one grave after the other to find Yehuda's name. We still didn't find the grave and left the cemetery in disappointment. This ordeal was very strange to me and the only explanation I could think of was that this was a sign from Heaven that it was not the appropriate time to visit his grave.

The day before departure I went for a visit to the diamond and gem exchange center in Ramat Gan which is north of Tel Aviv. This is one of the biggest diamond centers in the world. I met businessmen there from New York and some of them had come to visit their homeland like me. Among these people was a very rich person who sold emeralds from Colombia in New York. I had met him once or twice and this time, he remembered me and was excited to see me.

He initiated a conversation with me and said, "Listen to me carefully. If you work as an employee who gets a fixed salary, you will not advance far. If you want to make big money, you must be independent and that means being self-employed. I am not saying you should leave your place of work but you should have something else on the side. What's your opinion, if I leave with you a few packages of gems or a few large stones, and you sell them? Any profit will be yours. If you succeed, which I am sure you will, you will make nice money. I immediately answered, "Look, I'm not the guy for this. I work for the Ambuy Company and I cannot do this to my bosses." He would not take no for an answer and continued to convince me with other tactics. He asked me if I had a work agreement with my employer and what was written in it. I answered, "Nothing is written in it because there is no agreement. My boss said from the beginning that I am learning this trade like any other so I can help my family and brothers."

This man, who was over sixty and named Jack, continued to talk and try to convince me until I realized that he was somewhat right. He told me that when he saw me in New York, he was impressed with my politeness and speed and sincerely wanted to help me. We finished our meeting after I told him that I will think about his proposal and get back to him.

I went home and told Dad about this episode. Dad, who was a very practical man and had foresight, said to me, "I've known your boss well for many years. If you accept the proposal of this man, Meir will not be angry at you at all. He is aware that everyone who starts like you wants to ultimately be self-employed. This is natural. Don't be scared to check out this proposal more thoroughly, since Jack only wants to help you."

A wonderful month of vacation elapsed and I separated from my family again, and from their beautiful country. Upon returning to the office in New York, the whole staff had missed me and was waiting for me. They were happy to see me again. Someone said, "Michael, this office cannot function without you. We deeply felt your absence." Upon my return, I was already thinking about marriage.

I consulted with my bosses about getting married. They responded very enthusiastically and immediately approved a special vacation to accommodate this event. Together with them, I determined to have the wedding at the end of this summer.

Now, I had to make a proposal to Dahlia. I wasn't sure she was not ready for it, even though I knew she really wanted it. She was already twenty-three years old and this summer she would be twenty-four. The plain truth was that neither of us had our parents nearby to help us with our expenses and endeavors for the wedding. We had to prepare everything ourselves.

Dahlia's and I visited her sister Molly's house for a dinner. I announced my intentions to marry for Dahlia and we talked about the wedding. Her husband Dave, a typical American, was very happy about this and he naturally wanted to be sure that Dahlia had found a good match. He had a long discussion with me and was impressed. He asked me to determine the date for the wedding one month after his pregnant wife would give birth to the twins she carried. Molly had already lived in New York for ten years and had been married for three years. This birth was going to be the first for her.

The four of us then determined the specific date which would be Sunday, the twenty first of August, 1977. The next day, I announced our decision first to my parents over the phone and then to my bosses at work. Dahlia and I had a bit of savings from work and we set aside all of this money for the wedding. We reserved a catering hall called 'Regency House' on Jamaica Avenue where most of the members of the Jewish community in Queens held their events. This was a strictly kosher catering hall. Now both of us had to work harder and for more hours since the expenses were going to be tremendous.

Jack, the emerald dealer I met in Israel, arrived again at our office with merchandise. My bosses were too busy and didn't have time for him, so he came to my room. He had merchandise from Columbia and he left it with me to sell. When we met, he always patted my shoulder to encourage me. He was happy to hear of my upcoming marriage and since he had to return home the next day to Hertzeliah Pituach, (an upscale luxurious suburb north of Tel Aviv) he invited Dahlia and I for dinner at a restaurant. This was the first time he met her and he told me afterwards that he was very impressed.

Now, I needed a place to live with Dahlia after we were married. My roommate Avi had just returned from Israel with his wife whom he recently married there. To my good fortune, he immediately found an appropriate apartment with one bedroom near our building and moved there. I then decided that my current apartment would be quite good for Dahlia and I. I threw out all things that I didn't need, painted the whole apartment, and bought new furniture for the bedroom.

On a Sunday of July 1977, I invited Dahlia, Molly, and Molly's husband to show them the apartment I had prepared. It was a hot day and because of the power demands, there were blackouts all over New York. To my good fortune, the guests decided to stay in my apartment for two days since I had power. I gave Molly and her husband the bedroom, Dahlia slept on the sofa, and I slept on the carpet. Even though our bed was quite comfortable, it was difficult for Molly to sleep because she was late in her pregnancy with twins. A few days later, she gave birth to a boy named Benji and a girl named Sharon. It was cause for a huge celebration, due to Molly and her husband having children at such a relatively old age.

A month was left before our wedding and I sent two round-trip tickets to my parents. Dahlia sent one round-trip ticket to her mother so she could come. Her Dad was not able to leave his work to attend the wedding. Many preparations were necessary like hiring a rabbi, two kosher witnesses, an orchestra, photographers, and much more. Fortunately, I had close Israeli friends whom I became acquainted with in New York. They helped me a lot in the preparations. I had a friend named Gideon was a professional singer. He volunteered to perform at my wedding and recruit an orchestra of his friends who would perform free of charge.

Another Israeli friend approached me and said, "Michael, I want to save you money for photographers. My brother and I are amateur photographers and I promise you that we could do our jobs like professionals." Of course, I immediately agreed to his offer. I went to the local rabbi of the Sephardic Jewish community, Rabbi Shalom Hecht, whose office was close to my home and registered for his services.

We went to City Hall to register ourselves as a new couple to receive a New York State marriage license. My boss Meir invited Dahlia to his office to give her a gold ring, in which he added a 1.08 karat diamond. This was a present for her. I bought her a golden wedding band for the ceremony with my own money. My bosses had to go on an urgent trip to their office in Thailand but promised us that they will be back before the wedding.

My parents arrived from Israel. My sister and I drove to Kennedy airport to pick them up. I was sending invitations to all of our friends and acquaintances in New York as well as to all the friends of my parents from Afghanistan who resided in New York. I knew that my parents meeting their old friends at the wedding would enhance the celebration. Altogether, we invited three hundred guests. I wanted my wedding to be modest, so I decided to use my car instead of a limousine.

A week before the wedding, my bosses were still on their business trip away from New York. Dad accompanied me to work day after day that week. He was curious to see what I do. During my lunch break, Dad and I went to a restaurant together and went home earlier than usual, because I had to make the final preparations for the wedding. My Father became acquainted with more people that week and asked me to invite them too.

It is customary in the Jewish religion not to see the bride in the last few days before the wedding. Dahlia was staying at her sister Molly's house with her mother who had already arrived from Israel.

The wedding was on Sunday, the seventh of the month of Elul in the Jewish year 5738. A barber came to my home that morning to prepare me for the wedding. I rented a beige tuxedo with a brown bowtie. Friends of mine helped dress me. My car was ready outside with a friend as a chauffeur and my photographer friends were ready to take pictures and a video. The invitations stated that the ceremony would start at five p.m. Nevertheless, Dahlia, close relatives and I were already at the catering hall at three o'clock.

Dahlia, my bride was in the makeup room and a professional was taking care of her. One of the photographers would run between us to take pictures. My father, sister, and brother-in-law passed the tables to make sure everything was ready. Mom was excited as usual and I tried to keep my cool.

Around 4:30, the guests began to arrive a few at a time. The local rabbi along with a rabbi from Israel, and two *eidim* asked me to sit down with them for the writing of the Ketuba. This document is actually a promissory note by the groom to the bride to provide for her in the style to which she is accustomed, with a prenuptial agreement. This is the Jewish custom, that the husband will pay her a specific amount in case of divorce.

Between the writing of the *Ketuba* before the *Chuppa* lasted for about half an hour, the guests enjoyed a smorgasbord with delicious food. When I left the meeting with the rabbis and mingled among the crowd, I saw Dahlia's Sister Rachel with her daughter Michal who came from Israel to attend our wedding. This was a real surprise because I didn't think they could come.

The time for the C*huppa* arrived and all of the people moved to an adjacent hall where a huge beautiful stage, covered with flowers and white decorations, welcomed the crowd. The rabbis found seats on the stage and the ceremony started. One rabbi made the special blessings for the new couple and the other read the *Ketuba* loudly. The ceremony ended with the breaking of a wine glass, which commemorates the destruction of the temple in Jerusalem. From that moment, we were married in the eyes of G-d and the crowd shouted "*Mazal tov! Mazal tov!*"

The orchestra began to play music while the groom and the bride left to the dining hall and the guests followed. The guests found their reserved seats and a huge variety of foods were served by waiters. My parents were amazed from the beautiful ceremony of the *Chuppa* which was well organized as well as from the food and the way it was served.

Half an hour after the meal started, the singer announced, "Ladies and gentlemen. Let's applaud for the new couple." While the crowd gave their cheers, we moved to the dance floor and most of the crowd joined us for dancing. Then Dave, Molly's husband, made a toast with a glass of champagne to honor the bride and groom. Dad blessed us in front of all the guests and my friends did their best to make the new couple happy. The people ate and danced alternatively until late into the night.

My new wife and I, and my very close relatives were the last to leave the wedding ceremony at two in the morning. Dahlia and I went to the Marriott hotel with all the gifts we received which were stored in two large sacks. We were too excited to fall asleep even though the room was comfortable and luxurious. We decided to open all of the envelopes with checks, and I did not forget to follow my parents' guidance to make a list of all the checks we received. The bottom line was that after all of our huge expenses for the wedding, we remained with a surplus of $4,700. Right after this calculation, we started to think what to buy with this extra money. The instant decision was to replace our old car with a new Chevrolet.

That whole week we were invited to the homes of our friends in the evenings, while in the daytime we made trips with our guests from Israel all over New York. On Saturday, we threw a party in the synagogue as well as at our home in accordance to Jewish law.

In the following months, friends in the Jewish community and at work talked over and over about our how beautiful and successful our wedding was. This made me very happy.

My Wedding, August 21, 1977. My wife Dahlia and I were married at the Regency House in Queens, NY. My parents and mother-in-law came from Israel to share in our Simcha.

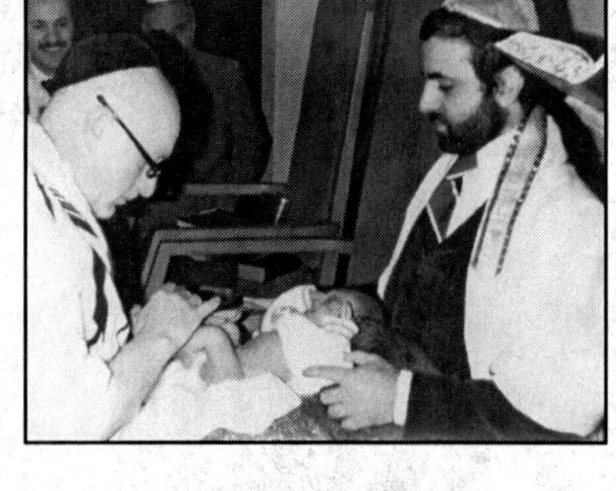

Top row: Our son, Yoni, was born soon after on February 13, 1979. I was the honorary Sandak at his Brit Milah.

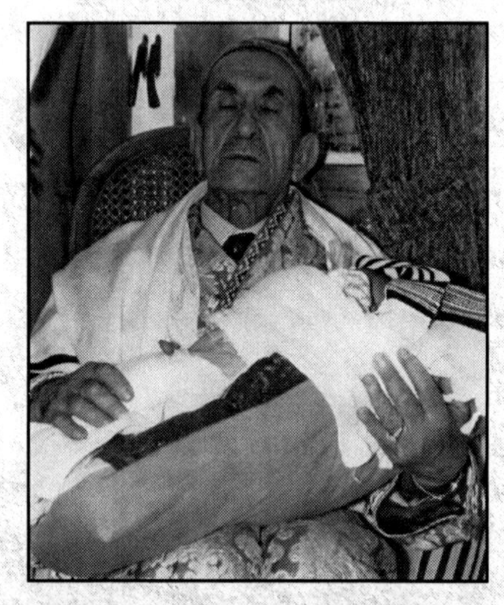

Right and below: Our second sone, Adam, was born on Election Day November 4, 1980. My father again came from Israel to be the honorary Sandak at the Brit Milah. Unfortunately, my mother-in-law passed away the same way Adam was born.

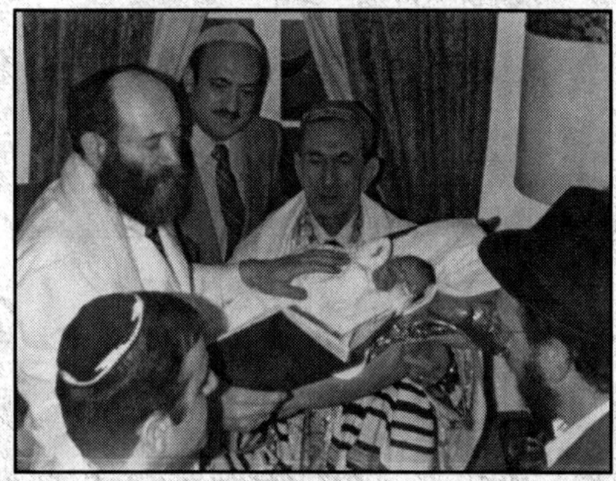

CHAPTER 28

After a ten day vacation from work, I returned to my usual activities on 47th street but with the essential difference of being married. Fortunately Jack the emerald dealer arrived at our office with plenty of merchandise. He also met my father and had a long conversation with him. At my apartment, Dad took me aside and said to me, "After speaking with Jack, I'm sure that his business offer for you is the chance of your lifetime. Be friends with him and cooperate with his business. Your bosses don't have time for you since they are too busy handling big business. You are not in their consideration. They didn't even come to your wedding, and for four years, you have been stuck with the same salary. I didn't see any signs that they were interested to give you a raise. Don't keep your head buried in sand."

My manager at work, *Mollah* Gavriel, noticed that I was very busy with Jack and he didn't approve of this. I succeeded to sell a few emeralds from the package Jack left with me quickly and gained a nice profit. It seems that there was slander in the gem market, since my emerald sales were reported in detail to my boss Yehuda, who had just returned from a business trip abroad.

My father was still in New York when Yehuda summoned me to his office. He said to me, "We heard that you have a business on the side and even opened your own business account. You cannot behave like this. We need you to seriously work full time with us. Gavriel, your manager, wants to leave here to be self-employed. We plan to replace him with you. To summarize my speech, I give you the option either to stay and work only for us or leave and be self-employed."

I was not prepared for such a speech. When I recovered my wits, I said to Yehuda, "I am not causing any damage to your business. Jack simply brings me merchandise and asks me to sell it." Yehuda shouted at me, "No! You cannot work for us both!"

The same day in the afternoon, I went with Dad to the bank to deposit a check in my business account. I was aware that my bosses didn't have bank accounts at this branch. Nevertheless, when Dad and I looked through the window, we were surprised to see Yehuda sitting with my bank officer. We immediately related this meeting to my earlier episode with Yehuda at his office and decided not to go in. It was a weird a moment when Dad and I stared at them, and they saw us.

Dad suggested that I talk with my other boss Meir, who had more authority in the company. Meir was still abroad and I had to wait for his return. The high holidays of *Rosh Hashana* and *Yom Kippur* came closer and my parents and the other Israeli guests of my wedding returned home.

When Meir returned to the office, I asked him for an appointment to discuss matters, which had been really bothering me. I told him everything in detail and he listened carefully. When I finished my story, he said to me, "I like you and I want you to advance but you must understand that while working with us you cannot work simultaneously for someone else. Maybe, G-d forbid, you would mistakenly sell one of our stones and think that it was yours. You will then deposit the money into the business account you recently opened." I immediately responded, "Sir, I am very thankful to you for all the years I worked with your company. However, I worked honestly and I would like you to check the company's inventory to prove this. I want you to do this in ensure that not a single gem is missing, before I leave to be self-employed."

Meir accepted my request and his secretary worked for two weeks to check the inventory. The results were that there were actually extra gems and not less as feared. I then resigned from my work place in a civil manner on good terms and my boss Meir gave me a wonderful reference letter. This recommendation helped me to be immediately admitted as a member of the DTA, the Diamond Trade Association.

For the first time in my life, I was totally self-employed and felt financially independent, free, and happy. I did not want to spend money to rent an office yet, so for the time being I worked from the offices of the DTA. I received all of the necessary services under one roof there, like a telephone, secretary, desk, and the means to set up meetings. My wife Dahlia continued to work as well and together we provided ourselves with a nice living.

One day, I decided to go visit my former work place, and the manager Gavriel welcomed me with yelling. He said, "I've slaved and worked here for twenty years and recently, I planned to leave and become self-employed. You left in a rush and because of you, I'm stuck here. My boss cannot find me a replacement. The plan was that you were supposed to replace me." I tried to calm him down and said to him, "If you remember well what happened recently, you are the real cause of my resignation. Besides, I am not keeping you here. If you want to leave so much then do so."

I continued to work from the Diamond Trade Association offices and from time to time, went out to make sales. I often made significant profits since everyone in the gem market knew and placed full trust in me.

The years of 1977 to 1980 were a flourishing period in the gem business. Many new companies were established, mainly by Afghans who learned the business from friends or relatives. They traveled to the Far East to buy gems, and then returned to New York to leave them with local dealers on consignation. I had to sell such merchandise, and if I didn't succeed, I would have to return it. I had between three to four months to pay these Afghan dealers for the merchandise if it sold.

It was 1978 and Dahlia and I decided to travel to Israel. I wanted my family to get to know my new wife. We went directly to my parents' home and stayed in a room they had prepared for us in advance. Jack invited us for dinner in his beautiful villa in Hertzeliah

Pituach and asked to have a meeting with me on the next day at the Diamond Exchange Center in Ramat Gan. I spent that whole day with him and learned many secrets of the precious stone business. He had plenty of experience and he motivated me to advance. That morning, Dahlia went shopping and in the evening she went to stay the night with her parents and four sisters.

One day, Dahlia and I decided to rent a van and take all of our close relatives for a trip to Jerusalem. I loved Jerusalem and could never tire of visiting there. Both of our parents attended the trip, with siblings, we were a total of fifteen people. We combed the whole city of Jerusalem including the Wailing Wall, the Arab market, and then the city of Hebron.

It was springtime and we enjoyed pleasant weather and blooming flowers. We spent the whole *Pesach* holiday in Israel and afterwards, Dahlia returned to New York. I stayed in Israel to purchase gems with Jack in the Diamond Exchange Center. Dahlia left a few pieces of her jewelry with me so that I could bring them back to New York safely.

On my return to New York, I passed all of the airport checkups and felt that I had one foot on American soil. At the last minute, one evil customs employee called from behind me, "Hey sir, come back!" I returned to his post and he led me into a nearby room to check my entire luggage. He opened the suitcase, looked inside it, and even tore its inner lining. Then he asked me to open my handbag and he did not find anything suspicious.

All of his searching did not placate his worries and he asked me to take off my pants and shirt. On my undershirt, I carried an attached wallet and inside it was a broach adorned with forty tiny diamonds. This was a jewelry item that I bought for Dahlia's birthday for four-hundred dollars. The customs official angrily asked what this was. I told him that it belongs to my wife who returned two days before me to New York after our trip to Israel. He disregarded this explanation and called his supervisor to report, "I just caught a diamond smuggler." I heard the voice of his supervisor on the other side of the line asking, "Diamonds? How many?" The customs official started to count the diamonds up to forty. To double-check, he counted again and reported back to his supervisor.

"Forty Diamonds?" his supervisor repeated. The customs official confirmed. "Bring him right away to my office." The customs official quickly closed my suitcase and instructed me to get dressed. He then read me my rights from a printed brochure, "You have the right to remain silent. You have the right to hire an attorney and counsel with him." I was in a state of shock and did not know whether to laugh or cry. Without warning, he placed handcuffs on me and called a police car which arrived at the rear of the terminal. He shoved me into the backseat of the police car and within a few minutes we arrived at the airport police station.

The time was late and I knew that Dahlia was anxiously waiting for me at home. Before leaving Israel, I called and told her not to bother to go to the airport since I would return home in a taxi by myself. At the police station, I had to fill out different forms and I had my mug shot taken as well as my fingerprints logged. I had the status of a suspected criminal. A police captain welcomed me with an unexpected smile. He was a really nice man with a good heart and understood the circumstances. He said to me, "You don't seem to be the criminal type. What happened here?" I answered him, "Sir, the confiscated item is a small piece of jewelry which I bought as a present for my wife for only four-hundred dollars in New York. We were on a trip to Israel and my wife returned a few days before me. I took it with me for safety. We recently got married and do not deserve to endure such an ordeal. Please check this item yourself and you will see that this is a mountain formed from a molehill." The officer then said to me, "It's too late for me to check it since it has been sealed as evidence. Nevertheless, I believe what you are saying and I am also concerned for your wife who is waiting at home. Usually in such cases, the suspect is not released until he is brought before the court but I will release you so you can go home. Tomorrow at exactly nine a.m., you have to appear in court." I thanked him profusely and left the police station.

I hailed a cab and arrived home. Dinner had already been on the table for a long time and Dahlia was very worried. We sat on the sofa together and I explained to her the ordeal I experienced. When she heard that I was handcuffed, she almost burst into tears. "What is going to happen now?" she asked. I said "I have no idea and I will only know in court tomorrow." Because of the stress, I was not able to sleep that night.

The next morning, I prayed a lot after getting out of bed. Dahlia went to work while I went to court. At the entrance to the courtroom, the evil customs officer who had arrested me was waiting. When he saw me, he immediately came close and invaded my personal space. There was a long line to see the judge and I waited for my turn with the customs officer behind me. Even when I went to the bathroom, he accompanied me. I was not summoned before the judge until three p.m.

He read the report of the customs officer and said to me, "It seems that you screwed up nicely and could end up in jail for this. Do you have a lawyer?" I answered, "No, your honor." He said, "Okay, then I am going to release you on bail for fifteen-thousand dollars and go to the next room to arrange your bail with the court officer." I went to this room as instructed and the officer asked me what assets I possessed. I told him that I had a few stocks and a new car. "What is the value of the car?" he asked. I answered, "About fifteen thousand dollars." He said that the car had to be left with the court and that I had to appear on a specific date six weeks from now with a lawyer. On my way out, I approached the customs officer and I gave him a glare that let him know that he had done me a great wrong. To my surprise, he understood my hint well and asked for my forgiveness. He said, "You have to understand me. I was a combat pilot and went through Hell in Vietnam. Afterwards, I remained unemployed and I thank G-d that the government was able to get me this job." I did not think that this was an excuse for his evil behavior and I continued to glare at him angrily.

I returned home and faced the problem of finding a lawyer. I never broke any laws, neither in Israel or here in America and therefore, had no experience with lawyers, and had no idea who to choose. Dahlia explained my problem to her coworkers. Her boss said to her, "Don't worry. Such things happen in America, and even worse. My husband is a lawyer and is experienced in similar cases. I will talk with him and he won't charge your husband so much.

An appointment was set for me with this lawyer quickly. His name was Tom Hecht. He had a big office with a lot of clients in Rockefeller Center close to my job. When I finished telling him the entire story, he burst out in laughter. He said, "This case is not as difficult as you imagine." He then asked me if I had a receipt of my purchase of the jewelry or a clear picture of the item. I told him that I had both. The lawyer instructed me to bring this evidence to his office and he would submit it to court. Two weeks after this conversation, I received a letter from court notifying me that this piece of jewelry had an appraised value of six-hundred and fifty dollars. This letter made me happy, but I still had to appear before a judge with my lawyer on the determined date.

The judgment day arrived and Mr. Hecht and I took a subway to court together. We entered the designated room and joined other defendants who were waiting for the judge's arrival. The judge came in and everyone present stood up. The first defendant was called before the judge and within one minute, received a sentence of three months in prison. Then the next defendant appeared and my heart was pounding after my lawyer was called out for a moment. He returned and asked me to accompany him outside. I asked him impatiently, "What happened?" My lawyer calmed me down and said, "You are free to go. The judge examined your case and dismissed all the charges." The lawyer was very happy for me, and said that in such a case, I could have easily ended up in prison even if I was innocent. New York is one big jungle.

A few months elapsed since my marriage to Dahlia and we got used to living together. The ordeal at the airport taught us both to be very cautious, since even by mistake, a person could fall great heights; especially in the diamond and gem business. The Jewish Afghan gem dealers made a lot of money and some bought villas in an upscale district of Queens called Jamaica Estates. As more and more Afghan Jews joined them, they decided to organize and establish a community center. The main initiators were the sons of *Mollah* Shalom Shamash, who were good friends of mine. One of these sons, named Weitzman, had a large house that he devoted the ground floor to being a temporary synagogue in the name of his father. This synagogue was named *Anshei Shalom*, people of peace. Since Weitzman and I had been very close friends in Afghanistan, he begged me to move to Jamaica Estates. He knew that I was a good cantor for Torah readings, and he needed my help. Unfortunately, we couldn't comply with his request since we couldn't financially afford the move. The houses in that neighborhood were tremendously expensive. However, the synagogue was opened and on Shabbat and holidays, twenty couples and their children attended the services.

Dahlia told me she was pregnant and due in February, and luckily her mother had stayed in New York since our wedding. She alternatively stayed with us or at Dahlia's sister. Dahlia chose a good physician in Manhattan who was affiliated with Lennox Hill Hospital. Every time she needed to be examined, we took our car early in the morning to this physician whose clinic was on 5th avenue, and from there we went to work. This was our routine until the winter of 1979. We started to buy a variety of things for our expected newborn. We didn't know the baby's gender yet, but we were very happy and didn't consider this relevant.

In the seventh month of Dahlia's pregnancy, we both had vacation from work for the upcoming Christian holidays and the New Year. I decided to make a quick visit to Israel since I heard that my older sister, Esther, was very sick. I didn't know what her malady was since my parents told me that this was such a rare disease that the doctors were unable to diagnose it properly so far.

I arrived at my parents' home in Israel and found them in a terrible mood. I heard for the first time that Esther had cancer and needed an urgent surgery. The next day, I went with my mother to visit Esther in the hospital. When we arrived, she was still in the operating room and we waited until the surgery was over. The surgeon came out first and told us that the surgery was successful, but she was still sleeping from the anesthetic and we were not able to see her that day. She would have had to stay in the hospital for a few days under supervision. Dahlia's sisters were also worried about Esther's condition and visited her in the hospital almost every day.

Finally, she returned home after a week but she was still in pain. One day, I bought her the best quality of fruit and pistachios and sat near her bed to talk. It was amazing to see that she was comforting me by saying, "Don't worry. First of all, you should take care of yourselves. You have a nice and brilliant wife and have just started a family." I had stayed in Israel for ten days and then returned to New York in a rather bad mood.

My wife, who saw my condition, tried to cheer me up and said, "I have pity for Esther. She's just thirty-seven years old with three children and she does not deserve this. What can we do if these things are a part of life? These are calculations of G-d and we can only pray for her."

One Friday, I called Esther's home to ask how she was doing. Her husband answered the phone and I immediately asked him about Esther's condition. There was an awkward moment of silence and I heard Dad's voice on the other line. He told me that Esther wasn't well and was taken again to the hospital. I asked him what he was doing at her home and what happened. He answered, "Nothing special. We just have to pray for her and everything will be okay." I did not know that about half an hour ago, she passed away and people were gathering at her home.

On Sunday morning, close friends and relatives visited my home uninvited. Dahlia's uncle, who was a pharmacist in New York, followed by her mother, came and started to set up the living room in expectation for more visitors. Then a friend of mine arrived, called me aside and said, "Sorry to tell you, but your sister has passed away. You have to sit *Shiva*."

My sister who lived in New York arrived in sobs and joined me for the *Shiva* in my home. During this terrible week of sitting *Shiva*, people came to comfort my sister and I. Friends of hers and of my wife served refreshments to the visitors. The rabbi of the Jewish Afghan community in New York, Tzvi Simantov, also visited and encouraged us.

The day of our child's birth was approaching and Dahlia and I were getting more excited with each passing day. On the Monday morning of February thirteenth of 1979, Dahlia had an appointment with her physician. We took our car as usual and reached the physician's clinic. He examined Dahlia and determined that she had to immediately go to the hospital to deliver the baby and he will follow us. I returned to the car and drove like a crazy man to Lennox Hill hospital in Manhattan.

Dahlia was admitted to the hospital immediately and I waited nearby for her physician's arrival. She was already suffering from contractions every five minutes, and I didn't know what to do. The physician arrived and asked me if I was entering the delivery room to attend the birth. I answered in surprise, "Me? Oh, no!" Then he said to me, "But your wife told me earlier that she wanted you to be with her when she gives birth." I felt that I had no choice and agreed to enter.

A nurse approached and gave me scrubs and a cap to put on and told me to wash my hands with soap. I entered the delivery room with the physician and immediately heard screams from my wife and shouted orders from a nurse, "Push… Don't push…" After a few minutes, the baby was out and Dahlia's physician said, "Congratulations! You have a baby boy!" I instantly looked at my wife and she was suddenly quiet. The nurse cleaned the baby and brought him to his mother. Dahlia was still in too much pain to hold him, so the nurse gave him to me and I got very excited. He was a cute little boy but cold to the touch. I said to the nurse, "He's very cold." She replied that it was normal. After Dahlia relaxed, the baby was brought back to her while I stepped out.

I went straight to a payphone and called Molly and Dahlia's mother. I told them the good news and they rejoiced. Right after, I called my parents in Israel to give them some happiness after the tragedy of losing a daughter. I asked the grandparents of my child to come for the *Brit* and offered each of the grandfathers the honor of being the *Sandak*. Unfortunately, both of them refused since it was too difficult for them to travel at the time. I decided to be the *Sandak* of the child myself.

I had eight days to prepare for the *Brit*. I hired a catering company who would serve the guests at the synagogue where I prayed. On the day of the *Brit*, we woke early in the morning and dressed the baby in warm clothes because it was snowing outside. We entered our car and I drove very slowly to the synagogue due to the heavy snow. All of our friends, acquaintances, and relatives in New York attended the ceremony. At the circumcision ritual, we gave a Jewish first and middle name to the newborn as is customary: Yonatan Yaakov, or in English, Jonathan Jacob. For short we called him Yoni. We chose the name before the boy was born, or before that we knew it was a boy. We had only prepared names for boys but not for girls.

Dahlia decided to stop working to stay at home and dedicate her time to raising Yoni. As our firstborn child he received a lot of attention. Often in the middle of the night, Dahlia had to wake up and nurse him. We took pictures and videos of him nonstop, and bought him only the best quality things. We were so happy that we celebrated his birth date every month.

I was self-employed in my business and I was very busy with travel to Israel. I always met Jack on these trips in the Diamond Exchange Center in Ramat Gan. I bought gems there and Jack added new emeralds which he bought in Columbia every few months. These emeralds were extremely high quality. Jack used to encourage me and gave a boost to my confidence. Since Yoni was born, I experienced a prosperous time in my business. My interpretation of this was that he was a blessing.

Slowly but surely, more of my Afghan friends moved to Jamaica Estates. Whenever I saw them, they said, "Michael, we miss you here." Their congregation had grown larger and the current synagogue which Weitzman donated was too small. One day, Weitzman called and said, "We are looking for a building for the new synagogue. We are establishing a community like we had in Kabul. This is for the future of our children. I am asking you again to think it over and to move to Jamaica Estates." I said to him, "You know that I want to move, but we still have to save money to afford it. Believe me that we will make it soon." Since it is forbidden to drive on

Shabbat and holidays and I lived too far from Jamaica Estates, I couldn't attend their synagogue services. Anyhow, on *Purim* when there were no driving restrictions, I joined them and was honored to recite the *Megillat Esther* with *Mollah* Gavriel Aharon in an Afghan melody. It was the first time that I attended a prayer service with a congregation comprised of only Afghan Jews in America.

The leaders of the congregation found a nice villa which was appropriate to house the synagogue. They made a quick negotiation with the owner and purchased the building for a good price. Right after this, in 1980, I saw a beautiful house right near this synagogue with a sign that said, 'For Sale'. I decided to try my luck and knock on the door. The owner was an American architect who designed his own house in a very professional way. I decided to bring my wife to evaluate the property since I needed a second opinion. She liked both the house and location. We bought it for a fair price of $110,000, partly a loan from the bank. I immediately joined the Jewish Afghan congregation of Jamaica Estates and became very active and popular in the community. We elected a sisterhood to represent the women of the congregation and they were in charge of organizing parties on holidays and after the Shabbat prayers. We celebrated the holidays in the synagogue exactly as we used to in Kabul. *Purim* and *Simchat Torah* were especially joyous, celebrated with dancing and singing.

All members of the congregation were very proud that finally we were united here in New York like one big family, as we used to be in Afghanistan.

On the first *Pesach* in our new home, we invited guests to join the *seder*. Among the guests was my friend and business partner, Jack who had arrived from Israel with his wife and his severely ill brother, Azaria. My wife invited her sister Molly, her husband Dave, and their twins who were already two-and-a-half years old. The beautiful and joyous *seder* in our customs from Afghanistan lasted until midnight.

Azaria came to New York to have a kidney transplant in a Manhattan hospital. Until a compatible kidney would be found, he stayed in a rented Manhattan apartment and underwent dialysis treatment in the hospital. All of Azaria's expenses were paid for by his brother Jack. Because of his business, Jack and his wife had to return to Israel. One day, I received from him a phone call from Israel and he sounded very excited. He said to me, "We found a kidney and it is on its way to New York. Please go to Kennedy Airport and meet someone who will arrive on a flight with the organ. When you get it from him, get to the hospital as fast as you can and leave it with the doctors." I said to him, "Don't worry. I'll take care of it." The next afternoon, I delivered the kidney and Azaria's transplant surgery was successfully. Fortunately, his condition improved completely.

Jack returned to New York from Columbia with more merchandise. This time, I had my own office with two rooms on 48th street between 5th and 6th avenues. Jack was very happy to see how successful I was doing, and was glad there was privacy. He poured all of his stones on the table. I had never seen such beautiful emeralds in all of my life. The high quality, color, and purity were unusual. Such stones are very expensive, and sometimes are worth more than diamonds. Jack told me that he didn't think that these stones would sell in New York because of their high value, but would sell very well in Paris. I responded, "What would you lose if we first tried to sell them in New York? If we do not succeed, then we can go to Paris and continue from there to Israel." Fortunately, the big buyers of emeralds in New York were a few Afghan brothers whom I knew very well. They were tough businessmen but good buyers. Jack and I went to them and to his surprise, they bought many stones, but not the most expensive ones. We decided to sell these stones on the European market, which Jack was thoroughly familiar with.

Two days later, we bought two tickets to Israel via Paris. Jack knew Paris well and spoke fluent French. He also held proficiently in Polish, Romanian, Russian, Spanish, English, and Hebrew. He was a very wise man, and also a shrewd businessman. It was a great pleasure to work with him and I learned a lot. We landed in Paris in the afternoon and reserved two rooms at a hotel. We planned to start work the next morning since we had scheduled appointments with two customers. Early the next morning, the receptionist's voice said, "Bon Jour! Bon Jour!" as a wake-up call and I thanked him.

We took a taxi to a jewelry factory. The owner, an older woman who was also a jewelry model, welcomed us. Jack spoke French with her, which I did not understand a word of. Within fifteen minutes, she had bought the all of the expensive merchandise. Jack left her with the remaining jewels, so she could sell them for us. We left with empty-handed and we were happy with the profit. Jack said to me, "Now that we have successfully completed our business, I am going to give you a guided tour of Paris."

We took the subway, which was clean and quiet, to the area of the Eiffel Tower. We traveled the streets, visited the tower and the Arch of Triumph, and passed many quaint coffee houses. I considered Paris the most beautiful city in the world. We finished the tour when Jack became exhausted and we returned to the hotel. We took our suitcases, checked out, and went directly to the Charles De Gaulle airport to travel to Israel. We landed at Ben Gurion airport and went our separate ways. We scheduled to meet again after the weekend.

CHAPTER 29

Fifteen-thousand people now lived in Yahud and the town was featured in the Israeli headlines. Their representative, Mayor Chatoka was a very strong and energetic person. He spent the town's treasury on soccer and education. The town's soccer team rose to the top league. I felt a lot of pride because my older brother, Amnon, was the manager of the team. He was involved with soccer since his childhood.

Yahud signed a cooperative agreement with the American city of Atlanta. Since Atlanta was much richer and larger than Yahud, it donated the town millions of dollars every year. My brother introduced me to Mayor Chatoka as well as to the owner of the soccer team and school principals. Most of my time in Israel was dedicated to my family or in business with Jack at the Diamond Exchange Center. We assessed the situation there, and two weeks later I returned to New York.

Wonderful news welcomed me upon my return home. My wife Dahlia was pregnant again and our family was growing. My older son Yoni, who was already a year old, was a hyper and smart boy. I was able to fly in my parents to attend the birth of my second child. I knew that this would make them very happy. I hoped that we would have another boy so Dad could be the *Sandak*.

I became more involved with emeralds and rubies at work. The ruby market was in high demand and their prices rose accordingly. I was acquainted with a new Polish gem buyer, named Pollack. I made a few good deals with him until one day, he offered me a job working with him. His offer really flattered me, but I said to him, "I'll need more information before making any decisions." He responded by inviting me to his office.

His office was opposite Rockefeller Center on the top floor of a luxurious building. We had the meeting with his secretary, who wrote down everything we said. He offered me a salary of $50,000 a year, plus partnership for colored stones. He said, "All colored stones I purchase for the office will be sold by you. The partnership will be 50/50." I answered," It sounds good, but I have to consult with a close friend before accepting."

I spoke with Jack, who knew Pollack, and he highly recommended that I take the offer. He then revealed to me, "Michael, you know that I am exclusively an emerald dealer, and the demand is low. So far, I have made a nice fortune and I am approaching retirement. I plan to decrease my business activity, and you will not make much money working with me. I encourage you to take the offer." This conversation opened my eyes to the situation I was in. I made a mistake of not signing written contracts with Jack. We went on trust, since he had known me from the time I was an employee of Meir and Yehuda.

Within one month, I closed my office, sold all the furniture, and moved to Pollack's office. I received my own office and started working immediately. Pollack was very busy and traveled most of the time. He purchased the gems and I sold them. One day, he decided to change the direction of his business and entered the field of toys. His office became a wholesaler for major department stores and sold huge quantities, resulting in a fortune of profits. I wanted to enter this part of the business too, but Pollack wouldn't let me. I stayed in the gemstone business with him.

I enjoyed being involved with our growing community as a temporary cantor. We needed more experienced cantors as well as other religious officials. Most importantly, we needed a good rabbi to be a spiritual leader who would guide and teach our youth. To our good fortune, a family of cantors joined the community. They were the Haimoff family who originated from Jerusalem, from the famous Bucharian neighborhood. This family lived in Jerusalem for generations and brought forth rabbis and cantors. The head of the family who joined our community was an excellent cantor with a tremendous and pleasant voice. His son Yigal was just like him. They were a family of good traits who enhanced the respectability of our young community and we were proud to have them.

Again the synagogue grew too small to accommodate the growing community and we looked for a larger place. The Hillcrest Jewish Center offered its building for sale, which had a large auditorium. The management of our community bought the building for a good price. The only disadvantage was that it was a bit too far from the center of our community, and it took me a twenty minute walk to get there from home. Right after relocating to this community center, we succeeded to find a young Afghan rabbi and cantor, named Rabbi Nasirov who had previously served as a cantor in Germany. We offered him the position of being our leader in New York, and within one month, his whole family moved to our neighborhood and rented a house next to the synagogue.

Near *Sukkos* of 1980 my parents arrived in New York and planned to stay with us until just after the birth of our second child, which was expected at the beginning of November. My brother Amnon called often to update me on the successes of the Yahud soccer team. He happily reported that the team succeeded to win many games and was then among the top soccer teams of the State of Israel. With the money Mayor Chatoka transferred to the team, they could hire the best coach in the country to train them. Besides that, they also bought some of the best players to join their team. The new coach, named Zeltzer, was one of the highest paid coaches in Israel, and was worth every penny. Since there was plenty of money for the team, the coach promised the players bonuses and higher salaries if they won.

Over that summer, the Yahud team came to the United States to play American teams. The team was accompanied by the coach, his wife, the mayor and my brother Amnon. Two weeks before their arrival, my brother notified me to be prepared to show them hospitality. In the Hebrew newspapers in New York, there were many articles about the team and my brother. The Israeli consulate provided them with security personnel to protect the entourage from any surprises.

I was sitting in my office as usual and on my break, I read in the paper that the team had already arrived in New York and were staying in a local Manhattan hotel. While reading the article, I received a phone call from one of the best players on the Yahud team, the goalkeeper, named Chaviv. He told me that they were already here for three days and needed my help. I asked him what their issues were, and he answered, "We have plenty of problems and we need you to come immediately."

The hotel wasn't far from my office and I drove over immediately. All of the team's players wandered the lobby in confusion. I asked Chaviv what was going on, and he told me that all of their plans and events were canceled due to security concerns. The Israeli consulate in New York did not want to help. Mayor Chatoka was very nervous and out of reach. He locked himself in his hotel room and wouldn't talk to anyone. Chaviv thought that I might have been able to do something to help. I calmed him down and promised to try my best.

Chaviv showed me the mayor's room and I knocked on the door a few times. The mayor shouted, "I do not want to see anyone! Leave me alone!" I shouted back, "Don't you remember me, your honor? I am Michael Cohen from Yahud!" The mayor was happy to hear my voice. "Ah, Michael, come in!" he said and immediately opened the door. I entered and he was bitter and tough. "This is your America? You should be ashamed of yourselves! We are here for three days and the press wrote a lot about our visit, and suddenly, we had our events and games canceled for no good reason. Our soccer team had won the national cup this year and is an icon in Israel and look at the disrespect we receive here." I interrupted him and said, "Your honor, calm down. I will take care of you and a few players this evening. I'll take you to a restaurant and for a tour of the city. The players who have relatives in New York can visit them." He answered, "Mr. Cohen, this is a perfect idea to save an unfortunate situation." We then parted.

That evening, I picked up all the players that could fit in my car and I drove to a luxurious Kosher restaurant named Moshe Peking for dinner. We stayed there for almost two hours. Only the mayor and I drank liquor, since the players weren't allowed to drink. Chatoka enjoyed drinking heavily and he was no longer bothered by whether or not the team would play. He accepted that in America, there are people who would not keep their word, and he would not fight with them.

Saturday evening, I returned to the hotel with a friend and an extra car, so the whole entourage from Yahud could accompany us to an Israeli night club owned by Boaz Sharabi, a popular Israeli singer. He personally knew the mayor and invited the whole team to dinner at his expense. On Sunday, the team was invited by the American Soccer Association to watch a soccer game between the New York Cosmos and a team from England. This invitation was a compensation for what had happened. The stadium was completely crowded and the game was fascinating. The Israeli team was really excited, as that they had never seen such a high level of competition in a game before. This was the end of their visit in New York. From there, they moved on to Philadelphia and played there. After what they endured in New York, they were not in shape to play and lost the game. From there, they returned to Kennedy Airport where I personally came to bid farewell to Chatoka. He was very thankful to me and invited me to come to his bureau on my next visit to Israel. He said, "I want to see you, and any assistance you need, I will provide." I responded, "Yes, your honor. I plan to come to Israel next month and I promise to visit you. I know that you are very busy and hope that you will have time for me." He promised to always be able to find time for me and even to let me borrow his official Volvo. At the end of this exciting conversation, he gave me a book on the town of Yahud as a present. He autographed the cover and we parted.

A month later, I indeed traveled to Israel. My parents and Amnon welcomed me with special respect saying, "Don't ask how thankful the mayor is to you and is expecting you in his office. We request that you accept his invitation and visit him. On this occasion, invite him to our home to spend an evening with us and experience our Afghan cuisine."

The next morning, before going to the Diamond Exchange Center, I decided to pass by the mayor's bureau. When his secretary heard my name, she immediately said, "Mr. Cohen, you are welcome to enter his office." The mayor welcomed me very warmly and we sat together to talk for a short while. He invited me, his closest staff, and a few players that had a close relationship with my family for dinner that evening in one of Yahud's restaurants. The secretary immediately called the restaurant's manager to reserve the restaurant just for our arrival, closing it to the public.

That evening I approached his home, not far from his office, and realized that he lived in a beautiful villa in a yard surrounded by walls. The front gate was open, and on the road stood a number of protesters. I didn't know what they were protesting, nor was I interested. I entered through the gate, rang the doorbell, and the mayor welcomed me warmly, but he was very nervous and stressed. I immediately said to him, "Your honor, it seems to me that this is not an appropriate time to go out for dinner. Maybe we can schedule for another time." He responded, "No, no, Mr. Cohen! What are you talking about? The people outside are protesting over nonsense. I will invite their representatives to my office tomorrow and deal with their grievances. This is something very easy to solve." He picked up his phone and called the head of the demonstrators on his cell to invite him to his office tomorrow to settle the matter. From his short conversation, I understood that these were young couples who demanded public assistance for housing since they did not have enough money. They claimed that the mayor did not do enough to help them, and transferred the public funds to the soccer team. I heard the head of the demonstrators screaming that he wanted a meeting at that very moment. "No way!" the mayor shouted back. "I do not have time this evening. I have American guests with me now and I want to be a good host. Please do not disturb me now, and leave the area right away. Otherwise, I assure you that I will not deal with your grievances." He then hung up the phone.

He invited me to enter his official Volvo which he drove himself. I sat next him and we talked for about five minutes until we reached the restaurant, near the highway to Tel Aviv. Security guards guarded the entrance and the owner himself came out to welcome us. The staff and players were already seated inside waiting for us. We were served Middle Eastern cuisine and enjoyed ourselves until the late hours of the night. When we felt tired, we left the restaurant, but not before I invited the mayor to dinner at my parents' house for the next evening, and he accepted the invitation.

When I returned home, my family was asleep. When I woke up the next morning, my father had already left for work, so I told Mom what occurred the previous evening with the mayor, and that he would visit us this evening. She said, "I am very happy for his visit and you know that I will prepare a perfect feast. Is anybody else coming?" I responded that I did not think so, but she and Dad could invite whoever they wanted to join the dinner. I also added, "Please don't make a big deal out of this, since the mayor is very modest and doesn't like to boast."

I left to Tel Aviv for business at the Diamond Exchange Center while my mother immediately started to make preparations for that night. I returned in the early evening and was glad to see a well-prepared table which had one end in the living room and the other on the porch. The weather felt pleasant and there was a beautiful view of blooming orange orchards. Under the building, a small river flowed with rainwater. Every night, the croaking of frogs was heard from this river and my parents resigned to live with the noise.

The mayor happily arrived accompanied by his son Chaim. Dad, the guests, and I sat at the end of the table and conversed. Mom came in from the kitchen and blessed the mayor for his wonderful achievements for the city and its citizens. He interrupted her speech, and asked about the noises he heard outside. Mom told him that this was the annoying croaks of frogs that occurred every night until morning, troubling their sleep. The mayor instantly responded, "Ms. Cohen, tonight will be the last that the frogs will celebrate. Tomorrow, there will be an end to this. I promise to take care of it."

I knew that the mayor enjoyed drinking so I poured him cup after cup of Israeli brandy. He really enjoyed this and started sharing confidential information about the city with me. As we say in Hebrew '*nichnas yayin yatza sod,*' in comes wine, out comes secrets. He told me that the money he receives from Atlanta is much more than the citizens are aware of. He immediately sent his son Chaim out to the car, to bring a letter with a current check of four and a half million dollars, signed by the treasurers of the city of Atlanta. I could not believe what I was seeing and asked, "What are you going to do with such a huge amount of money?" He told me, "I'm going to build a new high school in Yahud which will include a sports center and a stadium for the students. Besides, I am going to add more money for the soccer team which you know that I love very much. I intend to turn Yahud into a model city in Israel."

Mom served Afghan cuisine prepared with the main dish of *ash palow*. The mayor was so satisfied with the food that he insisted on getting the recipe from my mother. After we spent a few hours together, the mayor and his son thanked us and left our home. The next morning, we were woken by the noise of a tractor. Within a few hours, the tractor filled the little river with soil and cleaned the area. The frogs were heard no more.

After two weeks in Israel, I returned to New York. Yoni was very being spoiled by Molly, who assisted Dahlia voluntarily while I was away. She said to me, "Mike, your wife is tired and I think she deserves a vacation. I'll baby-sit for Yoni for the weekend, and you and Dahlia can travel somewhere to enjoy yourselves." I gladly accepted her offer and arranged for her to baby-sit for the following weekend.

Dahlia and I went to Atlantic City and stayed there in Hotel Glatt K from Thursday evening until Sunday afternoon. We had a great time in Atlantic City, but from the minute we arrived there, Dahlia wouldn't relax because she was worried about Yoni. She kept nagging me saying, "What is Yoni doing now? Is he sleeping? Is he eating?" I tried to calm her down by saying "Don't worry. He is in good hands." Before Shabbat started, we called home and discovered everything was okay. The hotel was occupied with many

Orthodox Jews and we joined them in the synagogue for the Friday evening services. Afterwards, all of us attended a celebratory dinner in the hotel.

The next afternoon, Dahlia and I wandered on the promenade parallel to the beach and as Shabbat ended, we immediately again called to check on Yoni. Everything was still fine at Molly's home. On Sunday, we left the hotel in a rush to return and see Yoni as soon as possible. When we arrived and saw Yoni again, it was as if we had been separated for a whole year. He was also extremely happy because he had missed us greatly. My wife then promised Yoni that we would never leave him alone again.

Dahlia was in her last months of pregnancy and our synagogue started to prepare itself for the upcoming high holidays of *Rosh Hashanah* and *Yom Kippur*. In a few weeks, my parents were expected to arrive and they would stay with us for *Sukkos*. I was sure that Dad would be happy to meet all of his old friends from Afghanistan and see our new Afghan Jewish congregation of young people which was growing steadily. All of us became stricter in observing the Shabbat and the holidays, due to being organized in a religious congregation under the guidance of a spiritual leader.

I drove to Kennedy Airport to welcome my parents and saw them in front of the terminal. They carried three suitcases and were looking for someone to pick them up. They seemed to be wondering why there was no-one was around to welcome them. I immediately stopped the car and helped them get in with their luggage. They looked at me as if I was the Messiah. "Where have you been?" they asked. "We have been waiting here for half an hour." I apologized and said, "I did not know that you would be ready to leave the airport so quickly. How was your flight?" They responded that this time, their flight was great and they did not encounter any difficulties.

We arrived at my home a short while later and my parents were very surprised to see my house. When they attended my wedding, Dahlia and I were still living in a one-bedroom apartment. Now we owned a huge four-bedroom house with a large kitchen and modern furniture. Our house was surrounded with a yard and had a garden where Yoni could play. "You have a beautiful house," they said and "We pray that you and your family live here happily." I responded, "Make yourselves at home. Ask for anything that you need without hesitation."

In the beginning, Yoni did not understand who these adult strangers were and I needed to explain to him that they were my parents and his grandparents: *sabba* and *savta*. It didn't take him long to get used to them, mainly to his grandmother because she spoiled him more. At night, he wanted to sleep in their bed.

My parents celebrated *Sukkos* and *Simchat Torah* with their Afghan custom of *Moed Bini*. They were very happy that we were a well-established community. Dad was especially proud because he was the one who encouraged me to go to America, and now he could see the status and respect I acquired within the community. One day while my parents were still here, I was approached by the leaders of the community who asked me to join them as a committee member and to be the community speaker. I declined their proposal and said that I was ready to voluntarily perform any function without need of a title. They tried to change my mind with different arguments but I insisted. Nevertheless, they did not give up and approached my father to convince him and succeeded. Dad came to me and said, "The community really needs help and organization and the leaders and I believe that you are fit for this job." When Dad asked me for anything, I could not decline out of respect and acknowledgment of his wisdom. From then on, I was involved in everything within the community as I was used to but this time, in an official capacity.

November arrived and I was on alert to take my wife to the hospital any day. We instructed my parents how to take care of Yoni in case we had to leave. On one Monday evening, Dahlia called her physician and reported pains to him. His response was that immediate hospitalization was necessary. I drove her to Lennox Hill Hospital in Manhattan that evening. Her physician arrived and again instructed me to follow him to the delivery room. I received a special uniform like last time so that I could be near Dahlia for the whole process of delivery.

Her labor began at two a.m. and two physicians were kept nearby. One of them was in charge of the delivery procedure and the other was a pediatrician. Lab tests that Dahlia took a few months ago showed that the fetus was not perfect. Another exam was requested, but since it was risky, we decided not to do it. We were decisive that under no circumstances an abortion would be performed since it was against our holy Torah. At the same time, we prayed to G-d to have a healthy child. The pediatrician was present to check the newborn the moment it appeared.

With G-d's help, my wife gave birth to a boy who was completely healthy and we were thankful to G-d for the gift he gave us. I was especially happy for the fact that my parents were here to celebrate with us after the terrible tragedy they had endured when my older sister passed away over a year ago. Dad was going to be the *Sandak* and I was happier for my parents than for myself. Right after the delivery, Dahlia was exhausted and wanted to sleep. The newborn was taken to a large room with other babies and I left the hospital to rush back home.

I wanted to tell my parents the good news as early as possible. It was four in the morning when I had arrived home and entered my parents' room. Mom was sitting on her bed while Dad was lying awake. In between them slept little Yoni. Mom asked immediately, "Is it a boy or a girl?" When Dad heard me answer that the child was a boy, he immediately stood up and both of my parents embraced and kissed me wishing me *Mazal tov*. They then started to interrogate me with specific questions. Was it easy? How were they feeling? How much does the baby weigh? And so on. My parents could not fall sleep due to excitement. I succeeded to take a short nap while I heard my parents discuss further plans.

In the morning, I called the rest of the family to notify them and to Dahlia's parents in Israel. I then went to the synagogue for the morning prayer and was welcomed very warmly by the congregants. They loudly wished me *Mazal tov*. I rushed to notify Dahlia's sister Molly of the birth, but she already knew about it from Dahlia. Dahlia also talked with her Mom in Israel from the hospital. I was worried about little Yoni. Who would take care of him until his mother returns? My Mom soothed my worries and said to me, "Don't worry about Yoni. I will take care of him in the best way and you don't have to miss work."

On Tuesday morning, November fourth, Ronald Reagan was elected presidency of the United States. My thoughts were on the fastest way to reach the hospital to see my wife and the newborn, which wasn't named yet. I arrived at the hospital and saw Dahlia holding and nursing our baby boy. She said to me, "What an enjoyment it is to feed him, he eats so nicely." I looked at him very carefully to see who he resembled. In my eyes, he didn't look like one of our family members. I wondered if he was our son or if there was a mistake in the hospital. I was aware that with time the baby's face could completely change. I spent an hour there until the physicians arrived, and then Dahlia's sister and cousin came to visit her. From the hospital, I went directly to work and the staff there already knew about the birth. Each of them blessed me and wished me *Mazal tov*.

Now I had to plan the *Brit*. I wanted to make it fancy, unlike Yoni's *Brit,* which was very modest since we were still mourning my sister's death. My parents were here and my father was going to be the *Sandak*, another good reason for the celebration to be fancy.

I could hardly concentrate at work and went home early. Mom had already prepared a beautiful table for dinner and all of us sat down to eat. Yoni, who noticed that his mother was missing and there was a special dinner prepared, felt that something important occurred. He jumped on me joyously yelling, "Daddy! Daddy!" I went to the kitchen cupboard and took out liquor with glasses to have cheers with my parents. Dad didn't change his customary behavior at the dinner table, even in my home in New York. Before the meal started, he brought a few cucumbers, peeled them, and ate them together with a cup of arak and we joined him. Mom then asked me if we had chosen a name for the baby yet. I responded, "It's good that you reminded me. I was so busy that I had forgotten all about it." I immediately opened a book of baby names, which I bought before

Yoni was born, and started to go through it alphabetically. The first name was Adam and I liked it. Mom immediately intervened and said that my father really hoped that we would name him Israel, after him. I didn't respond to her suggestion, but said to her, "Don't worry about the name, it will be alright."

Mom then served the meal. We hadn't finished eating, when I received a phone call from Molly. She said, "Mike. I have bad news. My mother passed away today in Israel, a short while after she heard your baby was born. She had a heart attack from the happiness and excitement of the news. She passed away on the way to the hospital." I was shocked and said, "It's impossible, Dahlia spoke to her after she gave birth." Molly started to cry and I tried to calm her down, saying that I would leave everything here and come to her home right away so we could think together about what to do, since Dahlia didn't know about the tragedy yet.

As she looked at me, Mom felt that something terrible had happened and asked me about it. I told my parents that Dahlia's mother had passed away and they were both shocked and went pale. I then apologized to Mom and said that I had to go immediately to Molly because she was not feeling well. I then said to her, "Please take care of Yoni," and left.

I drove quickly to the city over a Queens-Manhattan bridge and within twenty minutes, I reached Molly's home on the Upper West Side of Manhattan. Molly was crying hysterically, and asked for my forgiveness for not being happy for the newborn. She was mourning her mother's death, and asked me in sobs, "What should we do about telling Dahlia? She gave birth just a few hours ago. How could we tell her about this tragedy?" I told her that it was now too late and there were no visitation hours until morning. We then decided that the next morning, Molly and I would go to the hospital and consult with the physician in charge. I then returned home. Mom was awake and asked me, "What are you going to do now?" I answered, "We have no choice but to do what must be done: a *Brit* for our new baby on the eighth day after the birth." She continued, "But how will you tell Dahlia?" I saw that my mother was agitated and I said to her, "Mom, don't take this to heart. Everything that happens is from our merciful G-d. Tomorrow, I will take care of this matter, please go to bed now." She responded, "There is no chance that I will fall asleep. Tell me now what you are going to do tomorrow." I did not want to keep her awake too long and said, "Tomorrow morning, I'm driving with Molly to the physician in charge at the hospital to consult with him how to tell Dahlia the bad news. Good night Mom." I then left her room.

I went to bed with a bad feeling, and thought of how I could tell Dahlia that her mother was not alive anymore. She was young, only sixty-two years old. I was so exhausted I took a few short naps. From time to time, I heard my parents talking to each other while little Yoni slept in their bed between them as usual.

Wednesday morning, I went to the synagogue to pray as usual but in low spirits. Afterwards, I visited Molly and we went together to the hospital to meet the physician. He told us that Dahlia was doing well and could be discharged today. Regarding informing her of her mother's death, he suggested telling her about it at home. The physician then entered Dahlia's room with us and asked how she was feeling, and if she wanted to go home. Dahlia was not stupid, and saw on our faces that something was wrong, but said nothing. She told the physician that if his opinion was that she could go home, then she would. I immediately went to the office to take care of the paperwork and signatures, brought the car close to the building and we went home with the new baby.

On our way, I told Dahlia that her mother did not feel well recently. "What!" she answered, "I just spoke with her yesterday and she sounded healthy." I responded, "Yes, you're right, but afterwards, she didn't feel well because of the excitement." We arrived home and Mom had already prepared the necessities. Yoni ran to his mother to embrace her and noticed the new baby with us. He looked at him and did not understand what was going on. "Who is this?" he asked and continued, "I am the little boy at home, and I will always be the only boy at home." It was obvious that he was jealous from the sense of competition. We explained to him that this was his new brother and he had to love him.

People started to visit our home and Dahlia somehow felt that her mother was missing. We then told her the bitter truth. She did not cry loudly, and my mother warned her that she must control herself, since she had to breastfeed the baby. If she was stressed and nervous, it would not be healthy for the newborn. Dahlia totally accepted my mother's suggestion and remained as calm as possible.

On Saturday the religious rituals took place as usual. I was called to the Torah and candies were thrown all over in respect of the baby's father. When the Shabbat ended, I called a *Mohel* to schedule an appointment for Tuesday. The night before was the end of the *Shiva* for Dahlia's mother and therefore, we decided to have a decent celebrative party. The *Brit* would take place at our home and we arranged for a catering service to deliver food.

On Tuesday morning, many of our acquaintances from the synagogue and work visited our home to celebrate the *Brit* with us for our new baby boy. Among them were rabbis, friends of my parents, and other relatives. One surprise guest was Jack, the emerald dealer, who was in New York and heard about the party from our common friends at work. He came without invitation and I was very happy. All of these many people were condensed in our large living room and in the kitchen. The meal was served on tables in our huge basement.

Dad received the honor of being the *Sandak* and we named the boy Adam Israel. It was especially important for me to make my mother and father happy since they came a long way from Israel. Dahlia behaved extremely pleasantly during the celebration despite her deep sorrow over her mother's death. She was very strong throughout the ceremony and welcomed each of the guests warmly. The guests were aware of her special condition of grief mingled with joy, and they admired her behavior.

CHAPTER 30

My parents returned to Israel after five weeks in New York. They were so happy from what they saw in New York that they told all of our relatives and friends in Israel. They were especially proud of the status I achieved in the Diaspora in such a short time. My lavish lifestyle of a beautiful large house, a successful gem business, and a nice established family were sources of pride. Dad tried to convince my two brothers to join me in New York as he did with me a few years ago. Amnon, my older brother was still a bachelor. I really wanted him to come to New York so he could make a living and get married. The time for this was past due.

My younger brother Rafi worked in Israel in a gem polishing factory, a job that I had arranged. His expertise was with emeralds from Zambia. Dad offered Amnon and Rafi to go for just for a vacation in New York, exactly as I did. Both of them were convinced to at least try. They off from work and flew to New York. In our large home, we had vacant rooms so we easily and conveniently accommodated them. A few months after their arrival, Rafi told me that New York was not for him and he missed Israel and the mentality of the people there. We could not change his mind and he flew back to Israel.

Amnon couldn't decide if he wanted to stay in New York or go back. Dahlia and I kept him almost by force. We matched him with a Jewish Afghan woman who had also come from Israel to visit New York. Her name was Rachel Kashi and her parents were close friends of my family from when we lived in Afghanistan. Her grandfather was the righteous Jew who cured Amnon of his severe childhood illness with mysticism.

Amnon went out with her a few times and felt that she was his soul mate. He did not wait long to propose to her. Rachel's father was an old man, also visiting New York at that time, and he wasn't satisfied with the American-style proposal to his daughter. He insisted that the match would be done in the customary Jewish Afghan style. He really wanted his daughter to get married, and we did not want to give him a hard time. I complied with the man's wishes and came to his residence with a few prominent members of the Afghan community in New York to ask his permission for this match.

When we arrived, the bride-to-be was 'hiding' in her room. Her father asked why we had come to him and I said, "My brother Amnon wants to marry your daughter Rachel and we are here to ask for your consent." He responded, "It is not so simple. I have to ask my daughter if she wants to marry this man and if she agrees, we can go forward." This was the custom of Afghanistan and he wanted to follow it strictly. He rose from his seat and went to the next room where Rachel had been hidden. We waited impatiently for over five minutes until he returned and happily said, "She has approved and you have my consent. Now go and bring *shiriny*, and only then can we announce the wedding." We did as he asked, wished each other *Mazal tov*, and sat down to schedule a date for the wedding.

I called my parents to report the good news that Amnon would soon marry and promised to send them tickets to come. Dad checked with his workplace to see if he could receive a vacation, since he had recently traveled to New York. Unfortunately, his boss did not approve another vacation and only our mother was able to come to the wedding. Amnon did not have enough money for the expenses of the wedding, so I covered most of them. The family of the bride helped as much as they could. I was naturally happy to contribute my share knowing that this would help my older brother Amnon to get married and establish a family. It was a wonder, what a strange world it was, that Amnon and the girl had just arrived in New York separately for a trip and they were already a couple. Nothing occurred between them in the many years they lived in Israel and here, in a matter of days, they would marry. It seems that it was like our Sages say "*Meshaneh Makom Meshaneh Mazal*." A change of location is a change in fortune.

Amnon and Rachel had a wonderful wedding which all the Afghan Jews living in New York had attended. Right after, they traveled to Florida for their honeymoon, and when they returned, they decided to leave New York and reside in Israel. Rachel's elderly father lived in Israel alone in his own spacious house, and wanted the newly married couple to live with him. This house was in Kiryat Shalom, a suburb of Tel Aviv, where most of the Afghan Jews in Israel resided.

Our home was quiet again with no special events or guests. Yoni was already two years old and Adam was six months old. We decided to celebrate the upcoming *Pesach* in Israel. This was a nice time to travel the country since the weather was nice and flowers were blooming. After a great two week trip, I had to return to my work in New York. Dahlia and the children remained in Israel for a few more weeks.

Back at work, I was still as technically a partner of Mr. Pollack, but in reality, things were much different. He was busy with flights to the Far East, buying cheap costume jewelry, toys, and other items, and he sold them in New York. I was not involved at all in this aspect of his business, but I still received a monthly salary of $3,600.

On one occasion when I saw him in the office, I entered his room and talked to him openly. "Our arrangement is not working as we had settled. I closed my office and business in order to come and be your partner in the gem business. Now I see that you are dealing with totally different merchandise without my input. Why are you doing this?" He responded, "Yes Michael, I understand and I wanted to talk to you about this, but you approached me first. The field of gem trading that we worked together in partnership has stopped being profitable. It's not like it was in the past. I am a businessman first of all and therefore, profitability is the top priority. I am now doing business with merchandise that brings a high profit, when you deal with large quantities. I entered in a partnership with another company and there is no room for you to join. I am really sorry and ask for your forgiveness since we have to go separate ways."

I looked at him angrily since I helped him to flourish in the gem business for many years. I actually knew him from the time he was a plain laborer at a big company dealing in gems and diamonds. I helped him a lot during all those years, and it seemed that he had forgotten all of this. I did not argue with him but immediately ceased the partnership. I packed up my belongings and left.

I returned to self-employment in the same place I started in 1977 dealing with precious stones. I was aware that the time was different and difficult times lay ahead of me. I called my wife who was still enjoying her vacation in Israel and told her that my partnership with Pollack was over. I asked for her opinion on returning to live in Israel, since we had lived in New York for almost ten years. She did not let me finish my explanation and interrupted by saying, "No way! I am here in Israel and can see what is going on around. We have nothing to do here. Don't be so worried. You are still young and can succeed in self-employment like in the past. We will talk about this when I return." She then hung up.

Five minutes had not yet elapsed and my telephone rang. It was Molly who specially called to encourage me after Dahlia called her excitedly to report what had happened. Molly said to me, "I just heard from Dahlia that your partnership ended. You are a smart man and I am sure you will get out of these difficult times. A faithful Jew must believe that everything that happens to him is for his own good sake. It seems that something better is waiting for you. Don't even think about returning to Israel. It is difficult to live there and you have no chance to achieve the same lavish lifestyle you have in New York. I thanked her for her call and with no extra words, I stopped the conversation and hung up the phone.

I was home alone and poured myself a cup of whiskey. I had become calm and thought, 'What's going on with me? I have everything I could wish for and I am whining like a little boy.' At that moment, I decided to return to work to do what I know best:

dealing with emeralds. It was true that the gem business was slow, but it wouldn't necessarily stay that way. During this difficult time, I would cut down my expenses just to stay in the market and wait until the business picks up again.

At that time, many wealthy Jews from Iran fled their country because of the Islamic Revolution, and immigrated to the United States. They succeeded in smuggling their money out of Iran and invested it in real estate, diamonds, and gems. A luxurious community center was built in New York by Jews who came from the Iranian city of Mash'had. They established businesses here but preferred to deal amongst themselves. I tried to join them in their success by doing business with them, but I was not accepted. I was an outsider to them.

In Israel, the general situation got worse since the Israeli army got stuck in an endless conflict with Lebanon. They fought against Palestinian terrorists who found refuge in Lebanon and against a new Shiite militia called Hezbollah who were supplied with weapons by Iran. I received a phone call from a cousin of mine named Osnat. She told me that she was coming to visit New York in a few days. I welcomed her warmly like I did for every guest. This was a tradition in our family for many generations. It was springtime and the independence day of the state of Israel was drawing near. Osnat asked me if the Jewish community in New York celebrated this day. I told her that every Jewish community makes their own celebration. The Jewish Afghan community used to celebrate this day with a festive meal, music, and candle lightings. Our community rabbi led the ceremony, and I promised to take her to it. When the day arrived, I kept this promise and more. After the ceremony was over, we drove to Manhattan to another party which took place in an Israeli nightclub featuring a few of the top Israeli singers who came especially from Israel to perform for the holiday. The club was filled with Israelis who lived in New York and Osnat enjoyed herself.

One day, I received a phone call from my younger brother Ishai in Israel. He had bad news for me. Our cousin Erez, the son of our Aunt Bracha had been killed in Lebanon. I sobbed aloud when I heard this. Osnat heard me and came out from her room and asked me what had happened. I hung up the phone and told her the bad news. Both of us were in a state of shock and were paralyzed. She asked me for more details on what had happened but I did not know what to tell her. She then called her parents in Israel but they were not home to answer. I understood that the whole family had gone to comfort the parents. Erez had been a lively young man who would help anyone in need. He always had a smile on his face and it was a pleasure to sit and talk with him. I imagined to myself over and over what his parents, Bracha and Mishael, were going through. I gathered the courage to call their home but again, no one picked up the phone. This was a tremendous tragedy for the whole family and to the Jewish Afghan communities in Israel and in New York.

Our community rabbi here knew the family very well. Erez was a student of our rabbi when he studied in *Yeshiva*. On Shabbat, the rabbi eulogized Erez in front of the whole congregation who were deeply moved by his words. I personally could not overcome this tragedy and I was depressed for many days. Dahlia tried to make me feel better by appealing to my common sense. She said, "In life, there are happy moments and unfortunately, also sad ones. However, those who live have no choice but to continue." She also said to me that G-d does not want people to be sad because it means that they do not accept his decree, which is against a basic principle of the Jewish religion.

The newcomers from Mash'had established a new wealthy community in Queens, New York. Within a short time, they succeeded in doing business in New York and their community flourished. They bought beautiful houses for themselves and their quality of life was wonderful. The local media ran stories about them and they were frequently in the headlines. It was obvious that they had surpassed the Afghan Jewish community of New York, which was stagnant for years.

I was a community committee member and very active in my responsibilities. Nevertheless, my power was very limited. New members joined us constantly and we bought the building of the Reform synagogue to serve as our community center. This building was a half-hour walk from the homes of most of the members, which was rather inconvenient. However, we embraced this building for our necessities to accommodate our growing community.

Rabbi Nasirov did his best to advance his congregation but the inner politics of some powerful members hampered his efforts. I did not have enough power to assist him in his endeavors and it was a shame. Our little congregation had a president, two vice presidents, a treasurer, and a speaker. It was like managing a state. Each one of them thought that he had the ultimate authority and instructed the rabbi on what to do. They had no unanimous opinion and consequently their method of management failed. No progress had been seen in our community. Other members also contributed to this bad situation. They just came to the community center for a few hours on Shabbat as if it was a social club. This was not the way that we behaved in the cities of Kabul and Hirat, even though the conditions there were more difficult than ours today. I was aware that the world had been developed since the time of our forefathers, but I was still sure that our customs should not be compromised. This stagnation lasted for many years.

At work, I became acquainted with more people. One of them was a very honest person named Elchana who is originally from mashad and got into the emerald business in New York. We became close friends. He used to travel to Bogotá, the capital of Columbia to purchase merchandise. The emerald business was still slow but I asked if I could join him on his trips. He insisted that only when the market recovers would it be worth it to go with him.

It was 1983 and we were still struggling to stay afloat in the emerald business. The value of our home rose, so Dahlia and I decided to sell it and invest the money in a new business. We thought of opening a restaurant in Wall Street. A few potential buyers came to see our home and we decided to sell it to a Korean man, since he offered us the best price and paid in cash.

We bought a smaller apartment in a luxurious condo in order to invest the difference in business. We also had less maintenance expenses on the smaller house, but now in just a two bedroom apartment, we had to discard a lot of the junk that we kept in the old house. After a short while of living in the new apartment, we realized that it was quite suitable for us and our necessities. This building also had a swimming pool and gym for the tenants, which we enjoyed.

With a large sum of money in our pockets, the idea of opening a restaurant had become viable. Since I never did business in a rush, I went to consult with my Attorney Hecht, who had become a good friend of our family. He listened carefully to my idea of opening a restaurant and advised me not to do it. He told me that from experience with his clients, he would not suggest investing in a business that I had no experience in. In his view, I had to continue with my gem business which I had ten years experience with and knew many people.

I went to Elchanan's office one day and found him in a discussion with his father. He welcomed me warmly even though I arrived unexpectedly. I said to him, "I have a lot of money from selling my house and moving to a smaller one, and I want to join you on your trips to Columbia to buy emeralds. Elchanan's father, who was acquainted with this type of business, didn't even say a word. Elchanan said to me, "Michael, listen carefully. When I traveled to Columbia, the idea of taking someone with me never interested me. I didn't want to be responsible for the other guy, in case the business trip fails. However, I consider you a close friend and an honest person who sincerely wants to advance. I'll allow you to join me for a trip to Columbia and then you can decide. I will assign you a room in my Bogotá office and you will buy there whatever you want, and then sell it in New York for a profit." I left his office very happy with his proposal.

The next morning, I started arrangements for the trip to Columbia. I purchased a roundtrip ticket to Bogotá for a ten day stay. The date of departure was on a Sunday, two weeks before *Purim*.

I met Elchanan at Kennedy airport early in the morning. We were departing with American Airlines at nine a.m. I was excited since I had never been to South America before and was not acquainted with the place or the people. Anyhow, I felt confident because Elchanan was with me. From the minute we sat down on the plane, he gave me more details about our destination. He told me that Colombia was a poor country with a lot of thieves. You cannot walk with an expensive watch on your wrist or with a lot of money in your pocket. The people there would not hesitate to tear off a watch from your wrist. I told Elchanan that I would not walk in the streets by myself at all. He also warned me that it is easy to fall for business scams if you are not very cautious. We reached Miami and after one hour there, we changed to another airplane to Bogotá.

The plane landed in a relatively small airport in which I used my Israeli passport to pass through customs. The State of Israel had close relationships with Colombia because of military and agricultural cooperation. As an Israeli citizen, I did not need a visa to enter the country. When we left the terminal, a friend of Elchanan named Morris was waiting for us in his car. He helped us place our suitcases into the trunk and drove us directly to our hotel. We arrived at the famous Thikindama hotel lobby at five p.m. and entered the rooms we had reserved in advance. There were no kosher restaurants or groceries in the neighborhood of the hotel, which was in the old part of the city, so we ate food we brought from home. We spent the night in separate hotel rooms and scheduled to leave tomorrow morning and go by foot to Elchanan's office, which was a half-hour walk.

The weather in Bogotá is always pleasant. There is neither cold winter nor hot summer. It's like one long spring with temperatures in the range of seventy to seventy five degrees Fahrenheit and the leaves were always green. The city is on a tall mountain and it was sometimes hard to breathe. Colombia has special fruits with a rare taste as well as beautiful flowers and coffee plants. Coffee was the number one export merchandise of Colombia. This country also has natural treasures of emeralds that are world renowned for their high quality and beautiful color. Most of the mines are managed by private entities that pay a commission to the government. The mafia, which was very powerful in Colombia, was also involved in the emerald mines and businesses. Colombia is also known as a preferable vacation spot in South America. There are many resorts where tourists come for weekends and holidays.

I found a thriving Jewish community with wealthy Jews who had come from Europe to make business in Colombia. Most of them were in the textile business. They had two synagogues in Bogotá, one of which was close to our hotel. The more luxurious one was located in the new part of the city, a far distance from the hotel. The Jews of Bogotá had their own country club called Carmel where they used to meet every weekend. Elchanan suggested that it was worthwhile to visit it.

We arrived at Elchanan's office, located in a skyscraper. Morris welcomed us there. The office had a maid who offered tea and coffee to visitors. Morris was an Israeli with a sense of humor and always had a smile on his face. He kept kosher and supplied Elchanan and I with kosher food that he bought for himself.

We were sitting next to a large table, drinking coffee and waiting for merchants to do business with. Morris warned me in advance that this time, there is not much worthwhile merchandise on the market. He said," Don't buy a stone of low quality and always show Elchanan or me a stone before you purchase it. Usually, one tends to get excited the first time he does business here and consequently fails. I want you to buy something good here, and then you will return again and again." I responded to his advice saying that I understood, but he continued talking. "You don't know the Hispanic Colombians who offer you one stone and then switch with another and you would never notice."

The whole week, we sat and saw stones that merchants offered us but none of them were special or worth buying. Nevertheless, Elchanan was very patient and I followed his example. He told me that someone promised him to bring high quality gems in a few days because they were being polished. Meanwhile, Morris tried his best to be a good host. He prepared lunch for us daily, with a main course of Colombian fish. Each evening, he invited us to his house for dinner. To pass the time, we went out one evening to a beauty parlor. This was a place where rich tourists were spoiled with special treatments. We were welcomed with glasses of whiskey and offers for their services such as haircuts from highly skilled barbers or manicures. All of this was done together with fascinating stories with jokes and laughter. This whole extravagant service cost only ten dollars per person. In Colombia, this is ten thousand pesos which is considered to be a lot of money in a poor country.

By the end of the week, we had not succeeded in buying anything. On Friday evening, we went to the synagogue which was close to the hotel and occupied by *Ashkenazim*, Jews of European origin. It was an old synagogue built in the 1940's and the few congregants were elderly men. The next morning, we went to the synagogue again and this time there were more people. Since I was a guest, they honored me with *Maftir*. I made the blessings over the Torah in a modern Israeli accent which excited the congregants. Elchanan was proud of me. Right after the prayer service, we were invited to a *Kiddush* in the synagogue. We returned to the hotel at noontime to rest on Shabbat.

On Sunday, Elchanan and I were invited to the Carmel Jewish club. It was a beautiful day with clean air when we took a taxi ride over the high mountains to the club. The ride took half an hour and we enjoyed the view. The driver dropped us off at the gate, where a security guard checked us before letting us in. On the way into the club, we passed a huge garden and swimming pool.

Inside the building, we saw many people, mostly of middle and old age; they were seated around tables playing with cards and dice. They were so busy with their activities, that they hardly noticed our entrance. Suddenly, we heard someone loudly call for Elchanan. As we approached this man, he said to us, "I have been waiting for you for a long time now." He was a wealthy old man who spoke fluently in English, Yiddish, and Spanish. He invited us to sit near him and he spoke endlessly. It was obvious that he enjoyed talking, especially when others were listening to him. He offered us food, but Elchanan warned me not to touch anything except drinks. He quietly explained to me that anything barbecued here is not done with Kosher meat. We then went to the gym, the pool, and to the sauna and met people who did their best to enjoy life in this world. We talked with them about business possibilities and politics until we got tired and wanted to leave.

The rich man volunteered to take us back to our hotel in his car. On the way, he took a scenic route through the mountains where we saw the whole city of Bogotá. He drove through the new part of the city where the rich people lived in beautiful villas. We passed through a neighborhood where a group of young people stood, and our driver honked his horn at them in greetings. He slowed the car near the group to shake hands with one of them, and then we continued on our way. His face was full of pride while he asked Elchanan and me if we knew who those people were. He did not let us respond and he continued, "That was the son of Colombia's president." He dropped us off in front of the hotel and left.

Elchanan then told me that this Jewish man was one of the wealthiest people in Columbia and he owns assets that could enable luxurious living for many generations of his descendants. On Monday, we started a new work week with the intent of making good purchases on gems. We had to succeed in our mission by Wednesday night, since on Thursday we were scheduled to leave the country. Saturday night would be the beginning of *Purim* and I was eager to be home with my family. Monday and Tuesday passed with no business success. Elchanan became impatient and could no longer sit in the office waiting for merchants to come. He offered me to accompany him for a walk in the street hoping that we would miraculously meet people who would offer him high quality merchandise.

Outside, we passed unemployed people who gathered in small groups. We met a few gem dealers from Italy and New York who also complained that this time there was no good merchandise to acquire. We then entered a gem polishing factory and to our surprise, met Mr. Knoll, who was Jack's partner. He was sitting there and playing cards with a friend. When he saw me, he said, "Finally, you have arrived to Bogotá, but this is not a good time." I responded that in good times, he never offered for me to come. This response irritated

Mr. Knoll who instantly said, "You should tell Jack, not me. Besides, even now it isn't too late for you to succeed in Bogotá if you have a bit of luck." I wondered to myself and said to him, "Where is all the quality merchandise that Jack would have brought to New York?" He responded, "Yes, there were good times but now there is no raw material available. If there are some good gems available, they are very expensive." I asked him, "So what are you doing during these times?" He answered, "I just sit and play cards as you can see." He then rose from his seat, opened a drawer, took out a pistol, and put it in his pocket. He apologized for leaving because he had to go to a meeting.

Elchanan and I returned to our office while on the way, he told me why Mr. Knoll had to travel with a gun. "Mr. Knoll is an old man who could actually retire. He managed to make a lot of money in the gem business so he has to be cautious." I asked him, "What? Every rich person here carries a gun?" Elchanan said, "Yes. That is how it is. There is no state-sponsored security and people have to take care of themselves. Did you know that Knoll was kidnapped once?" I asked by who. "The Colombian Mafia," Elchanan answered. "They asked for a ransom of four million dollars in order to release him. There were long negotiations between his family and his kidnappers until they reached a compromise of one million dollars and he was released. From then on, he carries a gun in his pocket wherever he goes." I asked him, "And what about the police? Where are they?" He answered, "They are also afraid of the Mafia."

We reached the office and Morris welcomed us with the good news that he received a phone call that good merchandise was available and the merchants would come at two p.m. Elchanan was excited but said, "These Colombian bastards are purposefully coming in the afternoon when the sun is setting and the light is dim. I need to see the stones in the morning and I will not compromise. We will check the stones carefully and only finish tomorrow. Then we will give them a price offer."

We were sitting around a round table when a knock was heard on the door. Three short men came in. They were typical Hispanics and called us amigos while they patted our backs. The owner of the merchandise, who sat next to me, took a pistol out of his pocket and put it on the table. It was small but had a long barrel which was pointed at me. I gently moved it to another direction and they noticed that I was afraid. This caused them to burst out in laughter. I said, "Sorry but I have a weak heart and am afraid of guns." This made them laugh harder. Elchanan calmed me down saying, "Don't be afraid. They are friends." He chatted with them in Spanish for about ten minutes until they presented the merchandise. Each one of them took out a package of stones and poured it on the table. It was the highest quality merchandise that I had ever seen. They were colored a shiny, beautiful green and valued between three to ten karats each. I would roughly estimate at least one hundred karats in each package. I did not want to disturb Elchanan in his business and offered to go to the next room. Elchanan did not accept this and insisted that I should stay and watch how business is done. "But please do not tell anyone else what you see," he added. I was very curious to watch how the negotiations were done and how the merchandise was bought. The whole conversation was in Spanish. I could more or less tell what was going on by watching their facial expressions. Elchanan wanted to sort the gems and choose the ones he liked. This would take time and could not be finished that afternoon. This way, we would have to meet again the next morning so we could see the gems in daylight. They were interested in selling the whole merchandise without choosing specific gems, so the deal would come to a close on the spot. It was starting to get late and they insisted on getting a price offer for the whole merchandise. Elchanan gave them an offer but they did not agree to it. The end was that they accepted his demand to see the gems tomorrow morning. Almost everything was placed in a huge envelope, sealed with scotch tape, and signed by Elchanan. He asked to take only a few stones with him so he could check them at night in front of a light bulb.

We left the office exhausted but satisfied. We went to Morris' house for dinner. As usual, the meal started with soup and then fish. Suddenly, Elchanan stood up from his seat and took out the few gems he carried in his pocket and said, "I want to check them now before I decide to buy them tomorrow. They really looked terrific in front of a light bulb, shinier and stronger in color than when we saw them before in the office. Elchanan then said to Morris that it seemed to him that tomorrow we would buy many gems and maybe even all of the packages. Even though the deal had not been finished, we said L'chaim over Colombian liquor called Aqvartente which was similar to the Israeli arak but contained less alcohol. Elchanan then looked at me and said, "You have seen today how business is done in Colombia, but that is nothing compared to what you will see tomorrow. This will be a real battle with them but it will be resolved." Then Morris said to me, "It seems that you are not going to buy anything." I answered, "That is correct. The first time, you have to learn, but maybe tomorrow they will offer some special package for me. In any case, this was a good trip for me. I have learned a lot." We then told Morris that tomorrow we would return to New York to celebrate *Purim* with our families. We bid Morris goodnight and thanked him for this wonderful evening that he prepared for us.

Thursday morning was the start of *Ta'anit Esther* and all three of us were fasting but still at work in the office. At ten o' clock, the Hispanic gem dealers arrived again. Elchanan knew that the merchandise he was going to buy was good, and the Colombians knew it too. Both sides wanted to get the best price for themselves. Elchanan wanted to pay less and they demanded more. Elchanan also knew their character which demanded respect, and he gave them a lot. This was the best and quickest way to negotiate the price. Both parties shook hands, wished each other good luck, and the deal came to a satisfying conclusion.

As we were set to leave Colombia, Morris called a travel agency to reserve a flight for New York via Miami for that day. Since this reservation was done last minute, only one seat was available. We decided that Elchanan would take this flight and I would leave the next morning. There would still be time for us to arrive in New York before the eve of Shabbat. Elchanan preferred this arrangement since in his view, it would be dangerous to travel together with so many gems valued at millions of dollars.

Elchanan made final arrangements to leave to the airport and called a trustworthy stewardess, who had worked with his company for a long time, to come to the hotel. He gave her half of the gems he bought to pass over to Miami in order to not have any problems while leaving Colombia. He paid her a lot of money for this. In Miami, he would receive the merchandise and declare in customs that it came from Colombia.

Elchanan and the stewardess left for the airport and I remained in the office. An hour later, Morris returned and told me that tomorrow morning, both of us would leave to Miami with the rest of the gems and another trustworthy stewardess. He said, "From Miami to New York, you will carry this half of the gems." I was still fasting and had a few hours left for work. So far, I had not purchased anything.

A familiar couple entered the office with new merchandise to offer. They had already visited the office a few times but no business had been done. This time, they poured gems out on the table and they were exactly what I had been looking for. I saw fifteen to eighteen gems totaling about fifty karats. I wanted to pick the very best of them but they would not let me. I checked the merchandise very carefully and quickly decided to buy all of it. I then negotiated the price and closed the deal paying fifty-thousand dollars. Finally, I quickly returned to the hotel.

I packed all of my belongings since Morris told me that tomorrow at 8:30 a.m., he would come to pick me up to go to the airport. Everything proceeded like clockwork. We met the stewardess at the airport and gave her the gems. She hid them in her brassiere and all three of us boarded the plane to Miami which departed at 10:30a. m. Right after we arrived in Miami, Morris met up with someone in the airport who gave him the merchandise that the stewardess had smuggled. We then made a declaration at customs.

Now the plan was to take an American Airlines flight which would be departing from Miami at two p.m. to New York. When we approached the desk of the airliner, we ran into a serious problem. The scheduled flight had been delayed for two hours due to a technical problem. There was no chance then to arrive in New York before the eve of Shabbat. I was totally confused and did not know what to do. On Saturday night, *Purim* started with a recital of *Megillat Esther*. My congregation expected me to be there since I was supposed to read the *Megillah* publicly. Morris tried to save the situation by running from desk to desk to look for another available flight but with no success.

It was getting late. Morris came to encourage me and said, "If you don't want to violate the Shabbat, then you have no choice but to stay in Miami for the entirety of Saturday. Let's leave the airport immediately to a hotel otherwise we will be riding in a taxi on Shabbat." I accepted his offer and we arrived at the Saxony Glatt Kosher hotel in Miami Beach. Morris told me that he had uncles in Miami who lived close to the hotel, and he would spend the Shabbat with them. Before we separated in front of the hotel, we scheduled to meet again at Saturday evening in the hotel synagogue to attend the *Megillah* recital there. Afterwards, Morris would accompany me to the airport and I would continue my way to New York on a flight departing at 10:30 p.m. I approached the hotel receptionist while Morris left with the merchandise to his uncles.

I was assigned a room and went there quickly with my suitcase. I did not have time to take a shower but only to change into my Shabbat clothes. I then ran down to the synagogue on the ground floor and found the congregation in the middle of the *Kabalat Shabbat* prayer. I prayed with the hotel guests for half an hour and then we all entered the dining hall. It was crowded with Jews who had apparently come to spend the weekend in the hotel for *Purim* which would start tomorrow night. The dining manager welcomed me as a new guest and seated me at a table with two couples. During the meal, we introduced ourselves and had smooth conversations. It was a delicious meal in which I had ate meat for the first time in almost two weeks. I was exhausted from last week and especially from the last two days. I needed peace and tranquility. I did not stay for dessert and returned to my room. I took off my Shabbat clothes, got into bed, and slept until morning.

The Shabbat morning prayer started at nine o' clock and lasted until eleven. I then went to the dining room for a light breakfast during which I had another pleasant conversation with the people I had dinner with last night. I then went back to my room to take a rest until lunch. As I approached my room, I was shocked. The door I had locked was open and the entire room was a mess. All of my clothing was scattered all over the bed and nothing was left in my suitcase. I immediately checked the pockets of my pants where I had about $500, and it was missing. My wristwatch, which I had left on the table, was missing too.

I ran downstairs to the receptionist, leaving the door open as it was, and reported the burglary to him. He immediately called the hotel manager who asked me for details and if I wanted to call the police. I said to him that all of my valuables had been stolen and he had to do what was needed in such a case since he had experience. The manager told me to return to my room and wait for the police to arrive. I then had to think about what happened more thoroughly. Above all, I was in luck that the expensive merchandise was with Morris. He carried half of Elchanan's gem purchase and all of the stones I had bought for myself. I would have been in big trouble otherwise. I also thought that there was a possibility that these burglars knew that I had come from a business trip in Bogotá and expensive gems were with me. They had simply followed me to the hotel, perhaps? Now I was eager to finish this unfortunate visit in Miami and be back with my family for *Purim* in New York.

Two police officers arrived at my room and interrogated me thoroughly for an hour. They wrote down everything in a report, checked for fingerprints in the area, and gave me a copy of their report before leaving. I then waited for the end of the Shabbat to see Morris at the hotel. I had a headache from all that had happened. I tried to sleep to calm down but had no success. In the evening, I packed the scattered clothes into my suitcase and went down to the lobby to wait for Morris.

When the receptionist saw me passing by his desk, he called me over to him. He said, "You are leaving tonight and have to pay for your room." I stared at him and said, "You know that I don't have one penny left after the burglary. How could I pay you? I don't even know how to pay for the cab when I get to New York." The receptionist consulted with the manager over the phone about my situation and was given him the okay to let me leave without paying. The receptionist said to me, "I am giving you an invoice and please send us the money when you return home."

Morris arrived at the hotel and immediately noticed from the look on my face that something wrong happened. He asked, "How was your Shabbat? Did something bad happen?" I told him that my hotel room was burglarized and all of my valuables were stolen. He then asked me, "Can you tell me exactly what was stolen?" I responded, "All of my money was taken." "How much?" he asked. "I really don't know exactly but it was a few hundred dollars. The police came to investigate but they had no idea who did it. The news of it spread throughout the hotel and the guests are more cautious now."

The time of the *Megillah* recital was approaching, so Morris and I went to the hotel synagogue. I just wanted to listen to it as it is obligatory and then leave Miami immediately. In the synagogue, I asked Morris if the merchandise we purchased was with him. He answered, "Yes. I have two packages in my suit pockets. One is yours and the other is Elchanan's. He already called me and is expecting your arrival in New York." We had a few minutes left until a rabbi of Miami would start reciting the *Megillah* in the crowded synagogue for *Purim*.

We spoke until they started. I said, "Who needs all of this stress of smuggling gems from Columbia? You know that it is easy for criminals to follow us from the minute that we give out the merchandise to the stewardess and until our arrival to New York. This is very valuable merchandise and the 'professionals' could easily steal it. Why can't we send this legally and insured via a shipment company from Columbia to our office in New York?" Morris calmly explained to me, "Yes, you are right superficially. The problem is that to sending the merchandise out of Colombia legally costs a fortune, and therefore, dealers got used to the fact that there is no choice but to smuggle. This is part of this business."

At exactly eight o'clock, the rabbi started the recital which went quickly and finished after one hour. Morris immediately drove me to Miami's airport for the flight to New York. Right before leaving the car, he gave me the flight tickets and the two packages of gems. We bid each other farewell and he said, "Hopefully, we will meet again soon." I boarded the plane, which was only occupied to half of its capacity, and I sat in an unassigned seat where no one was near me. I was very cautious after what happened in Miami. My sense of responsibility was very high because of the millions of dollars in gems that I carried with me. From time to time, I took a nap but always checked if the packages were in my pocket. After awhile on the plane, I finally felt safe and enjoyed watching T.V.

The landing in New York occurred at about one a.m. I hoped that my wife would welcome me at the airport since I notified her of my exact time of arrival. It took me another half hour to get the suitcase and to leave the terminal. Outside in the darkness, only a few people were waiting and my wife was nowhere in site. Cabs were waiting in a line while a supervisor guided commuters into them, so I took the first one I saw. I lived very close to the airport and within fifteen minutes, arrived at my home. The building was well-secured and I felt safe and was able to breathe a sigh of relief.

I opened the front door quietly and found my whole family asleep. I did not want to wake up my wife but then I heard her whispering, "Hi, welcome back." She urged me to be quiet since the kids were still asleep and tomorrow morning we could talk. It was three in the morning and I had only four hours to sleep until I had to go to the synagogue for the morning recital of the *Megillah.* I felt very satisfied with the journey and happy to be home with my family. Nevertheless, it took me a long time to fall asleep since I was wound up from my experiences.

The alarm woke me and I drove to the synagogue of the Afghan community. Here, the *Purim* celebration reminded me of my past in Afghanistan. From the minute I entered the synagogue, friends welcomed me warmly and Rabbi Nasirov said to me, "We really missed you last night. It was difficult for us without you. Next time, don't make a business trip when it is near the time of a holiday or festivities." I listened to him and nodded.

The *Megillah* was recited in the Jewish Afghan melody which I always enjoyed. While I listened, I stood next to the rabbi as I did in previous years. After about an hour, the recital was over and as usual, the congregation sang a few *Purim* songs along with music. I felt totally different compared to last night in Miami where the children there made it difficult to hear the recital. Hopefully next year, I will listen to the *Megillah* recital in Jerusalem as we wish "*Leshana Ha'baah B'Yerushalayim*".

The business trip and its purchases were over and I now needed to sell the merchandise I brought with me. My plan was that if I could quickly sell the merchandise and make a good profit, I would return to Columbia. Elchanan's brother was waiting for me at my home when I returned from the synagogue. He was there to pick up the beautiful gems that his brother bought in Columbia that I smuggled.

I left for my office in Manhattan. I opened my package, poured out the gems, and looked at them in excitement. I thought to myself that they were so beautiful, but the market was slow then. It was not like the good times when I worked with Jack. I would need patience to wait for improvements in the market.

Wedding of my brother Amnon to a close family friend Rachel Kashi. Amnon came to America to see if he can succeed in the United States. Unfortunately, after his wedding in 1981 he decided to go back to Israel with his wife.

My family and I on our first trip together to Israel (Spring 1981).

My family and I in our new home in Jamaica Estates, NY in 1981. We moved to be a part of the new growing community of Afghan Jews in Jamaica Estates.

CHAPTER 31

My parents called to check if they could visit and stay with us for the upcoming *Pesach* holiday. "Of course," I answered my mother. "What kind of question is that? Even though we live in a smaller home, we will get along." Dahlia and I worked together to prepare for their visit. We organized the apartment and bought a lot of food that was Kosher for *Pesach*. For the *seder*, Dahlia invited her sister Molly and her children, as well as another brother. Along with my parents, there would be a total of eighteen participants for the meal. Since the apartment and the living room were small, we had to set the tables in a special structure so it could accommodate all of the guests conveniently. I thought it would be better if my parents would sleep in my sister's home, which was much bigger than ours.

When my parents arrived at my home for the *seder*, they were stunned to see my little apartment. Mom immediately said, "Oy vey! What happened to you? From a big, beautiful house, you moved to such a tiny apartment! It seems that your economical situation deteriorated." I calmed her down saying, "No Mom. You are wrong. Everything is okay. We simply had a lot of expenses for the business and needed money for them. However, our standard of living remains the same and our situation is excellent. Don't feel bad for us." The *seder* proceeded very nicely even though it was crowded. This actually contributed to a feeling of closeness among us. My wife knew how to be a good hostess and the guests enjoyed the evening and were thankful for it. As usual during these *seders,* Dad was quiet.

During the *Pesach* holiday, we received a surprising phone call from someone who only spoke in Afghan. He introduced himself to me as Fateh Chan Safi. My father was standing next to me and I quietly asked him, "Do you know who Fateh Chan is?" The instant Dad heard this name, he snatched the handset from me and said, "He's my old business partner from Afghanistan during the fifties and sixties. Mom quietly whispered to me that Fateh Chan's family were also our neighbors in Afghanistan and she was good friends with his wife. Dad talked with him very excitedly since he had not heard from Fateh Chan in nearly thirty years. When they finished their conversation, Dad gave me more details to who this man was. Safi was one of the richest people in Afghanistan and the number one businessman of imports and exports. He told my father that Afghanistan is in chaos after the Russians ruined everything. There were fights every day and people were running away. Safi succeeded to escape with his family and fortune and arrived in the States. Dad said that they were now staying in the Pan-Am hotel in New York which was not far away from my home, and Dad wanted to visit him that moment.

I drove Dad to the hotel and as we entered the lobby, we found Fateh Chan sitting there. When he saw Dad, he rose to hug and kiss him. This was a special Afghan tradition that involved three hugs. Afterwards, we all sat down together. Dad was curious to know how he found our telephone number in New York. Safi told him that he visited the diamond district in the city where he met a Simantov, an Afghan Jew. He gave him my telephone number and he called us after he found out that Dad was in New York.

Dad asked Safi how he was doing and he answered with whining. "We passed through hell, and those who are still in Afghanistan are in hell. Everything has been ruined by the Russians who have killed millions of our people and left another million *dar-badar*. Families with women and children are wandering in the streets with nowhere to go. The people have no weapons to defend themselves from the Russians, but despite that, they are fighting back. They are courageous Afghans who defend their homeland."

Dad and I sat for two hours listening to his horrible stories from the battlefield and did not interrupt him. When he finished his heartfelt speech, Dad told him that he sympathizes with the sorrow of the innocent Afghan people and added that only prayers could help them against such a heavily armed enemy.

Dad invited Fateh Chan and his family to our home as they did in Afghanistan. The Muslims in Afghanistan were in good relations with the Jewish community there and used to visit Jewish homes during *Pesach* to wish them a happy holiday. Dad remembered that Fateh Chan's family loved these visits very much. Fateh Chan thanked Dad for the invitation but told him that his wife and children would come but he could not. He was looking to buy a house in San Diego. However, he wanted to also check houses in New York and asked us for help.

On the last day of *Pesach*, my parents stayed at my home. The lobby security guard called us to report that a woman called Mrs. Safi and her son Chalid were in the lobby and wanted to visit us. Naturally, I immediately gave my consent to let them enter the elevator. We didn't know exactly when they would come, but prepared the apartment in advance for their visit.

Mom was very excited since the two women were close friends during the fifties and sixties and had not seen or heard from each other since then.

Mom went to the front door to open it when she heard the bell ring. At the entrance stood Mrs. Safi and her twenty-five-year-old son. They were dressed in authentic Afghan garments. Mrs. Safi's head was also covered with a large, wide shawl. The women kissed and hugged each other for a few minutes until they finally entered.

Dahlia and I welcomed the guests very nicely. Dahlia wondered how Mom and Mrs. Safi could recognize each other after thirty years, and still had such strong feelings of friendship for each other. We all sat together at the table for a special Afghan lunch of fish and *palow*.

During the meal, we had nice conversations in which Mrs. Safi was the main speaker. She said, among other things with tears in her eyes, "Afghanistan is not the same country which you knew when you lived there. Right after you left, everything turned upside down. Do you remember all the beautiful landscapes, good and cultured people that lived there? All of this disappeared with the war. *Kafar-ha.*" She continued to pour out her heart which was full of anger at the Russian invasion who left ruin and destruction behind.

Mom nodded her head with sorrow and said, "You are absolutely right. Such a peaceful nation does not deserve such a miserable destiny. I am happy that you succeeded to leave all of that misery behind and hopefully you will remain in the States and G-d will help you to become established. You are now in a free country and do not have to worry. America is a blessed country and therefore, people from all over the world come to live here. You should know that a place with a lot of water and rain is a blessed place. A country whose people are generous to the rest of the world receives plenty of water from G-d. You receive what you give."

Mrs. Safi let my Mom finish her encouragement and then said, "No, Mrs. Cohen, you do not understand us. You are right that the USA is a beautiful, free, and peaceful country but we miss our own state of Afghanistan. We are used to the mentality of the Afghan people and cannot find a common language with the Americans, even though we have lived here for almost a month. Do you remember how friendly the Afghans were to one another?"

My mother answered, "Of course Mrs. Safi. We left Kabul twenty years ago, and until now have still not met such friendly and open people like the Afghans. I still remember that we Jews lived with the Muslim majority like brothers and sisters. I hope and pray that the day will come when G-d will punish this evil Communist power, and peace will return to the land of Afghanistan." Mrs. Safi interrupted Mom and apologized for pouring out her heart. She said, "It's the first time since leaving Afghanistan that I have been able to express my feelings. This is because I consider you to be a very close friend as we were in the past."

Dad, who had become inpatient with this conversation tried to bring it to a positive end. He said to Mrs. Safi, "I am sure that we will be able to see better days in our beloved Afghanistan when the occupation ends and Kabul will return to being a paradise." Mrs. Safi then finished the conversation by saying, "*Een Shalah.*"

This visit lasted for a few hours and it was difficult for them to leave. Before leaving, they promised that the minute they have their home here, we would be invited to visit. My parents were very satisfied from this nostalgic reunion of old friends in a country distant from their origins. Chalid and his mother bid us farewell with hugs and thanked us for the nice welcome. They promised to visit us again.

When I was alone with my parents, Dad said to me, "Chalid is a nice man and wants to be your friend. You should accept his friendship and you could cooperate with him in business."

Their son Chalid and his brothers came to the States a few years ago and were in the carpet business. He asked me to remain in contact with him and his brothers so we could assist one another. He told me that my public school principal, Mr. Aqai Nadi, also fled from Afghanistan and lives in New York. To my surprise, he sells carpets on the weekends at the parking lot right near our building. Every Saturday and Sunday, it turns into a flea market where my ex-teacher has a booth.

I impatiently waited for Sunday to go and meet Mr. Nadi. I liked him when he was my teacher in elementary school and I was happy to hear that after I graduated, he had transferred to a high school named Lesai Habibya to become its principal. I knew that millions of Afghan Muslims who could flee after the Russian invasion did so. They spread all over the globe from neighboring Pakistan to India, and farther off to Europe and America. Most of them immigrated to America to New York, Maryland, San Diego, and Washington. In these places, they established small communities of Afghans. I met many of them in New York. They opened food stores here and were involved in the carpet business. Some of them were cab drivers and the more talented studied engineering or medicine.

I had passed the flea market every weekend but never entered. This Sunday, I entered enthusiastically to meet my teacher whom I had not seen for decades. I wondered if I would recognize him and if he would recognize me. Would he remember that he called me *Yehudi Bacha*? He was the teacher that when I was eight, picked me up along with some other schoolmates to go to the king's palace to welcome the American president of that time, Eisenhower. Would he remember this and that he gave me an American flag to hold and wave? I passed from one booth to another and carefully checked the carpet sellers. The place was crowded with potential customers who had come to find bargains.

I approached a carpet booth near the fence of the parking lot. The seller placed his merchandise on the fence as well as in his booth. I immediately noticed that these were Afghan carpets and I figured that this must be the place. I approached him and immediately recognized my ex-principal. Even though thirty years had passed, his face looked the same but older. I looked at him and said in Afghan, '*Salam Malek*.' He observed me thoroughly and then smiled from ear to ear. He said, "*Yehudi*?" We immediately shook hands and hugged three times. Both of us felt very happy to be reacquainted. Despite his customers at the time, he left them and dedicated a few minutes for me. I realized that he remained the same person as before with a strong sense of humor and a smile on his face. I did not want to disturb him in his business and said to him, "I see it is very hectic here and I would rather have a special meeting with you at another time." We exchanged addresses and telephone numbers and promised to be in touch.

When I left the flea market, I had mixed emotions. On one side, I was happy for this nostalgic meeting but on the other side, I felt sorry for the man as well as for the whole Afghan people. I thought to myself, "Amazing what a war can do! It causes destruction, ruins, and most tragic, homeless people. A homeland is your real home and then to have to flee it to save your life from a cruel war."

I heard more bad news from Mr. Nadi about what has been going in Afghanistan: a country that was once a paradise, especially during the fifties and sixties, turned into chaos. During those good times, we had a happy childhood and had all the best. All of the people there enjoyed freedom and were united like brothers. We Jews never heard or even felt the slightest sign of racism from the Muslim majority. Now millions of these good people are wandering from place to place and cannot find rest. They really did not deserve such misfortune.

Pesach ended and my parents really enjoyed staying with us. They liked the atmosphere of our home which reminded them of the days in Afghanistan. On *Isru Chag*, it was customary for Jews to go to great rabbis for a blessing. Mom told me, "You know that in New York lives the grand rabbi of Lubavitch. People from all over the world come to him to get a blessing. We are now in New York so we should take advantage of this opportunity. Maybe you could take us to him. This would be good for you too. It's never harmful to receive a blessing and I believe that it's good for your livelihood." I promised Mom that I would check how I could fulfill her request.

I never went to visit this rabbi and I never felt the urge to. Nevertheless, I would not object to accompany my parents who really wanted to see him. Many people from Israel, including leaders of the government, came specially to receive a blessing from this rabbi. They believed that he had a holy aura around him. Businessmen also used to come and consult with this rabbi, including many of my friends in the Jewish Afghan community of New York. I knew Mr. Sharbet who was in the gem business and went frequently to the rabbi for consultation and blessings. He was like a permanent resident of the rabbi's bureau in Brooklyn. I asked him to take my parents and me to the rabbi. He responded that on the upcoming Sunday, there would be a visitation period to see the rabbi and promised to come and pick us up.

On Sunday, he arrived at my home in his white Volvo and picked us up. He drove directly to the rabbi's synagogue in the Brooklyn neighborhood of Crown Heights. As we were accompanied by Mr. Sharbet, he guided us to a special path for VIP's and we quickly reached the rabbi. We passed many people who had been waiting on a long line. The rabbi gave each of us a newly printed American dollar along with a blessing for good luck. It was the first time that I had seen his face. While he gave me a dollar, he looked straight into my eyes and his face was shining. This sight made me tremble for a second. I could not talk to him or ask for a special blessing since his assistant stood near him and urged the crowd to move forward. Everything proceeded quickly so each of the hundreds of people who stood in line could pass in front of the rabbi to receive his blessing. Although the rabbi was a very old man, he stood for hours in his place to welcome each visitor.

We exited the building from the other end and Mom said to me, "It was worth coming just for the blessing. Don't you think so?" I answered, "Yes, I think I understand how blessings work. Nowadays, there are just a few truly righteous people and they really have the power to bless others. I know the saying of our ancient sages, *Tzadik Gozer Ve' Hashem Mekayem*, a righteous person makes a blessing and G-d brings it to fulfillment. I believe in it. I've heard of many people who came to this rabbi in the past and experienced miracles afterwards." I then turned to Dad and asked him his opinion of the visit. He responded that Mom was right and added that hopefully, the blessings of grand rabbis for world peace will come true and then it will be a pleasure for Jews to live in their homeland of Israel.

That week my parents returned to Israel after enjoying a whole month in New York. Before leaving, Mom gave me a special speech begging me to return with my family to live in Israel. "I wish all of my children to live close to me in Israel. You have been living in America for over ten years and I think that is enough for you. What do you have to look for in the Diaspora? We have a beautiful Jewish state and America is only good for visits. Jews have to live in their own country and be thankful to G-d that they have one."

In these regards, Dad had the opposite opinion. He said to me, "What does a young family have to gain in Israel? Just wars and worries. I highly suggest that you stay in America which is much better for you than Israel." I was also aware of the opinion of Dahlia's father which was similar to my mother's. He was a Zionist and an activist in the Israeli Labor party. He always told me that Israel is the best place for Jews to live. Dahlia and I had not yet come to a final conclusion on this matter. We left the decision for another time.

CHAPTER 32

The emerald is a very expensive stone which needs a high turnover to make good profits. The gem business was weak and all of the profits I made went to covering our expenses: private schools, insurances, apartment maintenance, and groceries. Many people left the gem business for real estate or restaurants. I did not have the courage or ability to switch to another business, as that I did not have any experience in anything but emeralds.

People knew me as an emerald dealer, which made it easier for me to buy and sell. The market deteriorated until it became almost dead. The rich people who were involved in the business could survive the difficult time. Nevertheless, they were willing to steal business from each other. No fairness remained and I could not compete with these types of businessmen. A few years ago, the situation was completely different. Trust and confidence were an integral part of the business. It was not like that anymore.

One day, my friend of mine Mark called me, requesting an expensive emerald of eight to ten carats. He had a potential client that was looking to buy such a stone and he was from out of town. Since I did not have such a stone in my stock, I called my friend David who belonged to my congregation and was also a gem dealer. He told me that he had a beautiful stone of eight-and-a-half carats and he wanted seventy thousand dollars for it. I immediately went to the gem exchange club and found David there playing *sheshbesh.* He gave me the stone and I took it to Mark's office. When Mark saw the emerald, he got so enthusiastic that he almost fell off his chair.

He whispered, "How clean! How beautiful! How shiny this stone is!" I told him that I wanted ten thousand dollars per carat for this stone. He started to argue with me about the price and finally said that his potential buyer was scheduled to come soon and then we would hear from him how much he was ready to pay.

Since completing this sale would involve three dealers, we needed to leave a margin of profit for each of us. As I realized that the buyer wasn't coming immediately, I left the stone with Mark together with a memo in which I decided to give the emerald away for seventy-five thousand dollars. I left Mark's office and returned to the club. An hour later, I received a phone call from Mark. He told me that the buyer had already visited him and they reached an agreement over the price. Mark offered to buy the stone from me for sixty-eight thousand dollars by check. I said, "Give me a few minutes and I will get back to you with an answer."

I wanted to ask the owner of the stone, David, for his opinion of the deal. I went to David and found him still playing his board game with a friend. I took him aside and did my best to convince him to lower the price. He asked me if Mark was going to pay with a check to cash on the same day. When I answered yes, he gave me the stone for sixty thousand dollars instead of the seventy thousand that he originally demanded from me. I immediately called Mark to report the good news that he could consider the deal to be closed. I then wished him good luck.

I asked him to prepare the check because I was on my way to pick it up, and Mark said to me, "No. Don't rush. Come tomorrow at ten in the morning with a receipt and you will get the check." I happily returned to David to notify him that tomorrow he would get from me a check of sixty thousand dollars. This completion of the deal would bring me a net profit of eight thousand dollars. Not bad for such difficult times in the business. We wished each other good luck and I went home. I came home very happy and Dahlia and I made a *L'chaim* for this successful business day.

That night, there was a fundraising dinner at the United Jewish Foundation (UJF). I was invited and saw there my other two partners for that day's deal, David and Mark. They were sitting at the same table with David's cousin who had the original ownership of the stone with him. Naturally, they were talking about the successful sale of the expensive emerald.

The next day as scheduled, I prepared a receipt for Mark and went to meet him in his office at ten in the morning. I would give him the receipt, and he would give me a check for sixty-eight thousand dollars. On my way, I saw David and his cousin entering the gem exchange club and surprisingly David turned to me and said, "I know who you sold the stone to. He is actually a friend of mine and we could do the deal between ourselves." I was stunned. "What? You mean that I am out of this deal?" David stammered and said, "Look. You do not understand, but you wanted to earn too much money here." I responded to him, "My earnings are none of your business! I succeeded to sell the stone while you were sitting and playing in the club." David and his cousin exchanged glances and laughed at me.

I tried to disregard this accidental encounter and ran to Mark's office with the intent to get the check as I had been promised. Mark said to me, "Sorry but the deal is not going to be closed with you." I was shocked and in front of him, I tore up the receipt I had prepared. I then asked to have the stone back and Mark surprised me again. "I'm sorry but the stone is not in the office and I will have it back tomorrow or the day after." This made me angry. I said to him, "This is how you treat a friend? Don't play games with me. Either I get a check from you or you return the stone." Mark tried to calm me down and said, "I don't think that you deserve an eight thousand dollar profit in such an easy deal. It's definitely too much." I responded that it seemed it was a waste of time to argue and I was going to complain before the arbitration committee of the club. "What you and David did is against the rules of the gem exchange club." I slammed the door behind me as I left.

As I was walking to the office of complaints, I could easily analyze what had happened. The night before at the UJF dinner, Mark, David, and his cousin were sitting together talking. They discussed the whole emerald sale process and the scenario apparently went as follows: Mark reported that he sold a stone to an American buyer for eighty-two thousand dollars. David, who listened to this news and asked, "Mark, may I ask how much you paid for this stone?" Mark told him, "Sixty-eight thousand dollars." David responded in astonishment, "What? This is my stone and I can sell it to you for just sixty thousand. Give back the stone and we can remove Michael from this deal." Mark understood that he would have more profit, and instantly agreed. When I went the next morning to meet Mark as planned, I did not know of this situation. All of this was done behind my back.

I went immediately to the office of complaints in the gem exchange club to fill out a complaint form detailing all of what had happened and requested arbitration. The officer arranged the arbitration for two o' clock on the same day and notified all of the parties involved. David and Mark arrived a few minutes before the scheduled time and as they saw me, they begged not to have the arbitration in the club in front of its president and committee members. They suggested to go to the office of David's cousin and to settle the dispute there. My instant response was, "No way. I do not accept this offer. I want the arbitration to be done in front of the committee members so you will feel ashamed." This wasn't characteristic of me, yet I felt this situation was so corrupt, the arbitration would prevent it from happening as often.

At the hearing before the committee, the respondents presented their version of the case first. One of the committee members asked me, "How much money do you want to receive instead of the lost profit and finish the case?" I said to the committee that I wanted the eight thousand dollars, as was agreed upon among the parties. The end of this hearing was a weird decision by the committee: since Mark wanted to give the check directly to the owner of the stone, I would receive five thousand dollars in cash on the spot and the case would be resolved. I felt that something was suspicious, but decided to accept the offer. I received the money in cash and left the room with mixed emotions.

A month later, I met David and he revealed surprising news. "Don't ask what happened with that whole saga. You are the only one that got a profit from the deal." I asked, "How could that happen?" David then got into details saying, "Mark sold the stone to that American buyer. He took the stone and immediately returned to his town where he filed for bankruptcy. All of his checks bounced and Mark did not see a single penny. Now he has difficulties paying us the debt." I felt sorry for what happened to my friends but I loudly responded, "There is a G-d!" I then said, "Why not earn money honestly and think of others?"

David apologized and asked for forgiveness and I considered the matter resolved. Nevertheless, it was a good lesson for me to remember. First of all, a person has to be cautious of even his closest friends. Second, the American precious stone market must be really down, and not something I wanted to be a part of, if people behave like this.

My business was in stagnation and Dahlia and I wondered what we should do. Her father realized our situation and told us, "As long as your children are small, it's the best time for you to return to Israel. Almost all of your family live there and it is a better place to raise the children." We sat down to consider if following his suggestion would be a good move. I felt that the process of leaving the States and going back to Israel would not be easy. I did not want to wander again as I did ten years ago. Dahlia was also not interested in returning, but was not decisive enough. Her father did his best to convince her to return to Israel. "What do you have to lose if you come just to try for one to two years? You already know Israel. It is not a new place for you and if you don't like it, you can always return to the States. You don't even have to sell your house. Just rent it out for the time."

Dahlia's father did not stop nagging her until she agreed to his suggestion but with one condition: to live in either Ramat Gan or Kfar Saba, the town where she was born. I told Dahlia that I was ready to travel to Israel to check these two places. However, I preferred to live in Ramat Gan since it's within walking distance of the Israeli Diamond Exchange Center where I planned to work.

In the summer of 1984, I notified my parents that I would soon come to Israel to check the possibility of return. Mom was overwhelmingly happy while Dad did not like this idea. I told him that I basically think like he did about my trip to America, and we should consider this return as just a long visit.

A week later, I arrived in Israel and after seeing my whole family, I went to a real estate agency in Ramat Gan to look for a proper place to live. The broker took me to show a few apartments until I made a final decision. It was a two-bedroom, beautifully furnished apartment near Bialik Street, one of the main streets in Ramat Gan. The monthly rent was three hundred and fifty dollars with a three-year lease. After securing a place for my family to live, I drove to the Diamond Exchange Center and rented an office in the gem section. A week after my arrival, I returned to New York.

New York was having difficult economic times, but the situation was even worse in Israel, where there was a recession and high unemployment. Despite this, the cost of living there was much cheaper with socialized healthcare and free private schools for the children.

My wife was happy that I made successful arrangements in Israel and she found proper schools for the children through her sister Rachel, who lived in Ramat Gan. As returning Israeli citizens who have lived abroad for many years, we were eligible to import electronics and furniture for personal use exempt from taxes. Dahlia and I rushed to department stores in New York to pick up electronics and furniture of the best brands and shipped them to Israel via an Israeli shipping company that used cargo ships that traveled between Israel and New York.

The date of return was scheduled for September second, 1984. We bought one-way tickets to Tel Aviv with the Israeli airliner, El Al. We then notified the congregation of our upcoming departure. The members and friends prepared us a farewell celebration at the synagogue's catering hall. It was an exciting event during which each of the speakers blessed us with good luck on the move. They said that we are taking a positive step and with G-d's help, they would follow us.

It was my last weekend in New York. I was filled with hope that G-d would continue to help my family as he always has. On Sunday, we would say, "See you later, beautiful New York," as we were leaving behind many good friends, positive adventures as well as harder times; and moving forward to another country. I would leave with mixed emotions, as it happened when I left Afghanistan, and again when I left Israel for New York.

My brother Ishai picked us up upon our arrival at Ben Gurion Airport. He turned on the radio, and I heard a familiar voice on the Kol Israel channel. Malachi Chizkia, my friend from my service, who spoke in a special language that we could not understand, ended up as a successful anchor for Israeli radio. I hoped to be reunited with everyone, family, friends, and even hoped to meet my commanders and sergeants from the army.

Whenever I return to the land of Israel, I feel proud for the privilege I had in serving the state of Israel, and being a part of it today. I always pray for the good sake of Israel, its continuing prosperity, and a future of peace.

On departure, I always wonder if I will return again to the place I was leaving. This time, I knew I would return to America. We knew the door to New York was always open and that we had a place to live. In each of our three countries, we experienced beautiful and enjoyable times which we will always cherish.

This was my journey among nations.

INDEX

Aliyah: Ascension

Amo: Uncle

Arava: Willow

Ashsava: A special, delicious food for Shabbat

Baq: Summer homes

Baq Omomi: National park

Bar Mitzvah: Considered an adult in regards to Jewish religious obligations, the thirteenth birthday for a boy

Barf Jangy: Throwing snowballs at each other

Barf pak: Clean the snow

Bazaar: Market

Boteh: Straw

Brit Milah or ***Brit Avraham***: Circumcision

Bro Bachim: Go, my son

Cha'a: A well

Chalifa: Teacher

Cham-Ula: Yeshivot, schools of learning Torah and Judaism

Chaparasti: Janitors

Choda Hamrat: May g-d be with you

Chevra Kadisha: Jewish funeral services

Chai chaneh: Tea-house

Ed Fatir: Happy Passover holiday

Een Shalah: With G-d's help

Erev Shabbat: Friday evening

Ethrog: The citron, a unique fruit of the citrus family

Farariha: Jewish Extradition

Gandana: A type of lettuce

Gil Sarsui: A natural hair shampoo

Gody Pran: Kite games

Gondy: Meatballs

Gvare: Baby bed

Hadas: Myrtle

Haftorah: Appropriate Biblical portion read on Shabbat

Hag'alat Keilim: Thorough utensil cleaning from any residues of Hametz

Haggadah: The story of the Jewish exodus from Egypt

Hamam: Bathhouse

Hametz: Food containing leavened bread or its byproducts

Havdalah: Ritual for the end of Shabbat

Havli: Neighborhood or Yard

Hojhin sahiv: Sir, formal

Hvale: Transfer

Idbid Bandan: Happy Sukka holiday

Isru Chag: The first day after a holiday

Jaz Chat: Permit

Kabalat Shabbat: Welcoming of the Shabbat

Kafarha: There is no law and no justice

Kalantar: Community leader

Ketuba: Religious marriage contract

Kiddush: Celebratory banquet

L'chaim: Cheers, 'to life'

Lary: Motor-vehicle for traveling

Leil Ha'Seder: Passover Eve ritual meal

Lulav: Closed leaf of a palm tree
Manda Nebashi: I hope that you did not get tired
Maror Maror: Boo boo
Mazal Tov: Good fortune, good luck, congratulations
Moed Bini: Holiday visits
Mohel: Circumciser
Mollah: Sir
Mollah Sahib: Honorable Rabbi
Motzei Shabbat: Conclusion of Shabbat on Saturday evening
Moza: Boots
Nan Vahi: Bakery
Palow: Persian-Afghan rice
Pana Bachoda: With G-d's help
Pareve: Not containing any meat or milk residue
Pasar Jan: My dear Son
Post Karakol: Furs
Qoda: Family relatives
Qomandon: Police commander
Rashi: Hebrew script style used in Sephardic countries named for the first rabbinical commentary to be printed in this style.
Rose sal: Blossom time
Rosh Hashana: Jewish New Year and judgment day
Sabze Gandom: Wheat
Sandak: Godfather
Sar Mahilim: Principal
Sarai shahazada: Commercial center
Sefre: Tablecloth
Shabbat Hagadol: The Saturday before Passover
Shafania: Small bunker
Shahar Kohna: Old City
Shahar Naw: New City

Shavuot: Holiday commemorating the receiving of the Torah on Mount Sinai

Shekem: An army canteen

Sheshbesh: Middle-Eastern backgammon game

Shiriny: Candy

Shiva: Seven days of mourning after the death of a close relative

Shochet: Ritual slaughterer for meat

Shofar: Ritual horn

Simchat-Torah: Holiday of rejoicing with the Torah

Sukkah: Ritual hut

Taqare: Washtub

Tasbe: prayer beads

Tefillat Haderech: Special prayer before leaving on a long journey

Tfillin: Phylacteries – Religious leather articles placed on the forehead and left hand during Morning Prayer in midweek after turning 13 years old

Tochom chori: Large traditional party

Toop Danda: A type of baseball

Tzadik: Righteous man

Yehudi Bacha: Jewish boy

Yom Kippur: Day of Atonement

Zamarod: Emeralds

ABOUT THE AUTHOR

Michael Cohen is a dynamic leader who serves as the president of a rapidly growing Jewish Sephardic community. He recently headed two Afghan-Jewish delegations in meetings with the democratically elected president of Afghanistan, Hamid Karzai. He helps to bridge religions and enhance American-Afghan-Israeli ties.

Michael, his wife Dahlia, their three sons, and grandchildren all reside in Queens, New York. His corporate office is located in the heart of the diamond district in Manhattan.

Look out for Michael's continued biography, coming soon!

Printed in the United States
113900LV00004B/94-102/P

9 781432 703295